Evolution Is Stupid!

EVOLUTION
IS
STUPID!

by John P. Verderame

GioiaMia
Prescott Valley, Arizona

Published by

GioiaMia
Prescott Valley, AZ

ISBN-13: 978-0615824130
ISBN-10: 0615824137

Expanded Edition June 2014

www.EvolutionIsStupid.com

For Laura and the Boys, who have evolved into fine young men

Preface to Expanded Edition

In May of 2014 my wife and I visited the Grand Canyon. It was my second time there, but my wife's first. The weather was spectacular, and so was the scenery, which no photographs can convey adequately, and it was fun to see my wife's reaction to first seeing it. No one can look at the Canyon without a sense of awe and profound thoughts about "how" it came to be.

We hiked the alleged "two billion year timeline" along the path leading to the Geology Museum, and took a few minutes to admire the scenery from the windows in the Museum.

As I was standing there (my wife had wandered off) observing, there was a park ranger seated before me, and a young man came up behind me to speak with him. He began to ask questions of the ranger, and this was the first of them, as we all looked down on the tiny green thread that is the Colorado making its way through the Canyon:

"Is there something special about that river that it was able to carve such a large canyon?"

My first thought was that the guy was a Creationist who was setting the ranger up and about to corner him, but it soon became clear that that wasn't the case at all. The man was simply observing the facts before him and questioning the conventional wisdom, and the ranger was on the spot.

The ranger gave a brief explanation as to how the Colorado river and its tributaries had carved the Grand Canyon, but the young man was not satisfied. He asked why other rivers generally don't carve canyons like that. He asked where all the strata had come from and how they had built up, only to have the river cut through them. He was genuinely perplexed that that little river could have done so much "damage."

Finally the ranger admitted that there is disagreement about how the Canyon formed, and left it at that. At that point I piped in and agreed that there certainly was disagreement, and gave my point of view. There was no further discussion.

But it was thrilling for me to see that someone was using his brain, observing the actual facts and processing them, and concluding that the conventional evolutionary long-age explanations were perhaps not adequate to explain what he was seeing. The young man disappeared quickly as I was speaking with the ranger and so I did not get to talk with him further though I wish I could have. If you're out there somewhere, KEEP ASKING QUESTIONS!

This expanded edition of *Evolution Is Stupid!* includes a section in the rear of the book with responses to feedback I've received over many years of dialogue with those who believe in Evolution and those who do not. Numerous topics are covered, not in depth but certainly in breadth, in the hopes they will be a stimulus to further study and discussion for anyone reading them.

My only desire is to get people thinking in the hopes that their world will be changed like mine was – for the better, and forever.

Contents

INTRODUCTION   1

    I Would Chuck Chuck If I Could Chuck Chuck

CHAPTER I   6

    Evodelusionists of the World Unite!

    You Have Nothing to Lose but The Chain of Being!

CHAPTER II   11

    Sex, Birds and Bees

CHAPTER III   15

    The Only Thing Evolving Is the Story Itself

CHAPTER IV   18

    Chuck Darwin Evolution Club Tactics 101

    Or, How to Intimidate Those Who Don't Believe Your Mythology

CHAPTER V   25

    Mysteries of the Universe Revealed!

CHAPTER VI   31

    Stupidity

CHAPTER VII   39

    Spinmeisters Extraordinaire

CHAPTER VIII   43

    SETI, or Signs of Evodelusionist Temporary Insanity

CHAPTER IX   46

    Come Together, Right Now, into Me

CHAPTER X   50

The Faith of Billions Can't Be Wrong!

CHAPTER XI   54
Don't You Love Spontaneity?

CHAPTER XII   58
Time Is On Our Side, Yes It Is!

CHAPTER XIII   62
Sex Again, and Something Fishy

CHAPTER XIV   67
Dinosaurs Take a Bite Out of Darwin

CHAPTER XV   72
Fossil Foolishness

CHAPTER XVI   77
A Body of Damaging Evidence

CHAPTER XVII   82
Millions and Billions and More Billions

CHAPTER XVIII   89
So, Are You a *REAL* Scientist?

CHAPTER XIX   95
Don't Know Why? Invent a Story!

CHAPTER XX   99
Proud to Be An Apeman

CHAPTER XXI   106
Not in *MY* School, You Don't!

Feedback Responses   109
For Further Reading   183
About the Author   187

INTRODUCTION

I Would Chuck Chuck If I Could Chuck Chuck

"When we consider, further, the freedom with which the writers of this school use their own imagination as an instrument of research – and the ease with which they construct vast edifices of conjecture on narrow and shifting foundations..."
--Giovanni Schiaparelli

While Schiaparelli was not referring to evolutionists in that quote, it is an accurate description of the way evolutionary research works, by a man whose own words were misinterpreted by "researchers" whose imagination flew way beyond the facts. You'll recall the story: Schiaparelli saw what he described in Italian as "*canali*" (channels – natural formations) on Mars, and the next thing you know those like Percival Lowell who had been influenced by the theory of evolution were seeing "canals" (made by Martians) and whole civilizations on the planet, yet without a single shred of hard evidence beyond their imaginations.

This is a book that should not even be necessary if common sense prevailed, but as the saying goes, "Common sense isn't too common nowadays," and I'm not the type who can sit on the sidelines and watch people be duped without saying something about it. I'm going to start out with the theme of the book, which will be repeated numerous times in the hope that perhaps in the end those who *have* been duped (or those who are *responsible for* duping others) will admit they have no answer for it, namely:

1

How did nothing become something and turn itself into everything?

Let me repeat that a bit louder this time, and you think about it.

HOW DID NOTHING BECOME SOMETHING AND TURN ITSELF INTO EVERYTHING?

Well, the answer to the question begins with a Big Bang:

Evolution is *STUPID*.

The whole evolutionary scenario from particles to people is STUPID and nothing more than the product of fertile imaginations. It's science FICTION, and it's about time people chuck Chuck Darwin and the whole Evolutionary myth. Evolution is a *religious belief* that has a grip on the world, and most of us can't seem to get out of the brainwash trap, or maybe we don't want to. Even though most people recognize in their heart of hearts that the whole evolution concept is stupid, they refuse to openly acknowledge it, are afraid to do so for fear of the wrath of their peers, and keep coming up with dumber and dumber imaginary stories that supposedly "prove" it, when it's simply not something that can be proven.

We should be LAUGHING at these dumb evolutionary explanations that supposedly tell us where we came from and who we are, instead of taking them seriously like we pretend to do. Yes, I mean *pretend*. Nobody could seriously take evolution seriously (you read that right). Your ancestors came from stardust that turned into bacteria that turned into fish that turned into monkeys that turned into you? Gimme a break! You don't believe that and you know it.

And those of us who know what a sham the whole thing is are always on the defensive. WHY? We have nothing to be defensive about. Why get defensive about a stupid concept with no basis in reality? We should be *laughing* the Evodelusionists into a corner where we can watch them squirm while they try to come up with the best and latest story to explain their myth. In case you forgot, all you need to ask them is:

How did nothing become something and turn itself into everything?

Then watch them get all mad and call you names and try to make YOU feel stupid and uneducated because you don't believe your great great great uncle was a tuna.

Every few months a new "pre-human" surfaces in the news, and the headlines scream that "THE HISTORY OF MAN WILL HAVE TO BE RE-WRITTEN!" Or some new fossil is discovered that "OVERTURNS!" our former ideas about how life supposedly evolved. Or astronomers find some new "evidence" in support of a universe that sprang out of nothing and made itself, and they add a billion years or subtract a billion years at whim. After all, who's counting? All we need to know is that the "experts" have assured us that we're all stardust from way back when, aren't we?

<u>Bull twaddle.</u> And it's about time people just come out and say so. I make no apologies for the toned down expletive. I live in the Wild West, USA, where there's a good bit of that stuff around, so I know it when I see it. Let that set the tone for the rest of this book, 'cause we're gonna do some evolution-bashing here. Sometimes I just can't believe people even take this stuff seriously. We should be MOCKING it. Your ancestors were fish? The "experts" told you so. And the "experts" are smarter than you. And you believe all that? No, you don't. So don't pretend that you do.

It's like a recent headline I saw proclaiming that a certain well-known astronomer had discovered "The Origin of Life"! So I read the article and what did it say? That we have the same elements in our bodies as are found in the stars, and that's where those elements came from. OH BOY, so THAT EXPLAINS EVERYTHING, now doesn't it? NOW I know where LIFE CAME FROM!! Right?

Wrong. For instance, it doesn't explain the following:

How did nothing become something and turn itself into everything?

Sound familiar? If not, it will soon.

****

A *caveat*:

This book is not for complacent or apathetic people, or for those who have impressed themselves with their great learning and degrees, so if that's you, don't bother reading it. I'm certainly all for learning and education, but I wrote this for humble people (learned and otherwise) who want to THINK about what they've been led to believe, and QUESTION the pack of lies called Evolution. It's for people who want out of the Chuck Darwin Evolution Club, but just can't do it, or are afraid to criticize or condemn what they know in their hearts is not true. It's for youths who are being indoctrinated, and not being allowed to think and formulate their own thoughts and opinions on the issue. It's for those who refuse to be told they came from slime that turned into fish that turned into chimpanzees that became people, and who know that's plain STUPID, but they're afraid to confront others who are insisting that either you believe what the Chuck Darwin Evolution Club tells you, or else you're an outcast of society, uneducated, or otherwise less evolved than they are. Yes, UNEDUCATED. That's what you are! Because you don't believe your ancestors were particles that turned into pond scum that turned into porgies that turned into primates that turned in to Peter and Patty.

■ ■

There's going to be a lot of sarcasm and some humor in this book, and I make no apologies for either. Neither do I make apologies for calling the promoters of the Religion of Evodelusion, those who think they have something to lose by abandoning it, "Chuckie Deeists," (think about it) who belong to the Chuckie Dee Evolution Club. That makes me think of W. C. Fields' "Chickadee," and many evolutionists are just chickens, and afraid to admit the whole thing is a sham because they fear peer pressure. I believe firmly that if they could chuck Chuck, they would chuck Chuck and everything that goes with him (or Him to Chuckie Deeists) including the whole Evolution myth, and if this book convinces a few people to stand up and speak out against what they KNOW is not true, then I've accomplished my purpose.

CHAPTER I

Evodelusionists of the World Unite!

You Have Nothing to Lose but The Chain of Being!

One thing you will hear repeatedly in the following pages is that regardless of what you believe about where everything came from, why it's here, and where it's going, my goal is that you be willing to step out and admit that EVOLUTION is stupid, and that it's about time we chuck it into the trash heaps of philosophy, history, biology, astronomy, geology and whatever other trash heaps we should toss it into. Because that's where it belongs. It's time to chuck Chuck Darwinism and Evolution and all its attendant lies, fables, stories, and stupidity. Stand up and say it's wrong! You know it is! And yes, I have solid beliefs on why I'm here, where I came from and where I'm going, but I'm not going to be cornered into defending myself. I'm gonna corner YOU. We're here to focus on *Evolution*, and that's what we're going to do. So if you're going to try to brand me as uneducated, or "one o' them creationists," or "not a scientist" or try to use any of the other tricks that Evodelusionist fanatics use to try to get out of having to explain themselves, it's not going to work. Why? Because I'm going to back you into a corner and slam you with the following question:

How did nothing become something and turn itself into everything?

Which you cannot answer and never will. However, just to be nice (you know, no bullying and all that) I will give you a little biographical information so you know what got me started on this campaign.

Back in the 1970s I was in college in Philadelphia, studying biology and other sciences – well, kind of studying if you knew me at the time. I was a confirmed evolutionist who actually used to have late-night arguments about how we evolved from apes and how certain sectors of humanity were less-evolved than others (I'll let you guess on that one – more on that to come).

Then, when I started actually THINKING about what my own college biology textbooks were saying, it began to hit me that evolution couldn't possibly be true. I mean, I was reading about a *"sudden explosion"* of complex forms of life in the so-called Cambrian period of geologic history. And about flowering plants *"suddenly appearing"* in the fossil record (from where? – nobody knew). And seeing *"evolutionary trees"* that showed an amoeba at the base of the tree, then fully-formed critters and humans at the ends of the tree branches with nothing in between, and those critters were fully-functioning, fully "evolved" organisms. "What happened to all the 'in-between' critters?" I wondered. Where did all the fully-formed critters come from, and how did they all "happen" to evolve just right to be able to work so well together (till humans came along, that is, and literally threw a monkey wrench into the mixture, or should I say "monkey wretch")?

Well, I could have comforted myself with the typical evolutionist response that "what we don't know now we'll eventually find out." But I thought, "Oh, really? Wouldn't that be more like 'FAITH' at work? Where's the 'Science' here?" What I then began to notice big time was that evolutionists invent faith-based stories to explain it all (they didn't see it happen, no one was there when it did, and they can't repeat it, so therefore yes, it's FAITH-based). And if the story sounds good enough, that's adequate. Doesn't matter if it's feasible, logical, testable, repeatable, or any of those other REAL SCIENCE thingies. If the story sounds good, that'll do. I began to see Evolution for more of what it truly is: a modernized, souped-up *religious myth* pretending to be science. So I was not going to just sit back and blindly believe it all. I questioned it, just like I questioned the religious system in which I was brought up as a kid. I dumped that too when I recognized the inconsistencies and unanswered questions inherent in the "system."

Now let's have some Fun with Evolution! I sincerely hope that some day you'll look back and LAUGH at the fact that you ever believed this stuff. But before we go, here's something more for you to think about…

***

Not long after the attacks of September 11, 2001, I was participating in a discussion board on the Internet concerning evolution. Of course, as an anti-evolutionist I was being mocked and ridiculed by Evodelusionist religious fanatics (we'll get into how evolution is really a religious belief later in the book), so I asked them,

"How many of you dusted off your *Origin of Species* and read it for a bit of comfort and hope after seeing the attacks of 9/11?"

I got no response (apart from a snide, meaningless one or two). NOBODY?? Well, IMAGINE THAT! You mean, the notion that you're a former scum ball that turned into a simian that turned into Sam the Man isn't a comfort to you? You mean that the belief that you're going to be dumped in the ground some day to rot and become fodder for future evolving things doesn't inspire you? You mean the fact that the perpetrators of the September 11 attacks along with every other doer of evil who ever existed (including yourself whether you'll admit it or not) are all just going or gone to the dust, never to be held accountable for what they did? Nobody who ever did evil – yourself included -- will ever have to answer for it? Doesn't that just throw a blanket of peaceful, easy feeling over the universe? Doesn't Evolution just give you that warm fuzzy feeling that there really IS no purpose to existence, no real reason for living, no tangible hope for the future, and that you and everyone else can do what you please and not be concerned about repercussions 'cause it's all part of "survival of the fittest?" Evolutionary Entitlement!! Doesn't that make you feel good??

No, it doesn't.

****

My website, EvolutionIsStupid.com, has been devoted to this topic since 2005, and has received hundreds of feedback emails. With the publication of this book the site is now dedicated mainly to feedback and additional information, and I encourage you to read it and join in if you wish. The site receives email from around the world, so please keep in mind that not all speak perfect English, and the letters are mostly unedited. I think you'll soon notice how often evolutionists try to divert the topic onto other things (more on that in Chapter IV), thinking that will free them from having to explain their OWN beliefs, and too how often the writers try to get off on the topics of God, the Bible, and Christianity, despite my insistence on trying to keep the focus on Evolution and Darwinism. Interesting, isn't it? The tactic is basically to try to place the burden on *me* to explain *myself*, rather than having to explain *their* beliefs, which include the following unanswerable question that you've heard before somewhere:

How did nothing become something and turn itself into everything?

Let's see if we can find out.

CHAPTER II

Sex, Birds and Bees

It all started out with the kissing bacteria. Yes, we all know that sex sells, and I want you to buy this book, so we're going to begin with sex cells, or sexy cells, or however you want to put it. My friend and I sat in a theater in a science museum – many years ago so I'm not certain, but I believe we were in Baltimore. And we were watching a movie about evolution, and it was showing how the sexes evolved.

So, there were these single-cell bacteria-looking organisms, you see, swimming around having a good time. They really seemed to be pretty happy, so who knows what prompted them to want to kiss, but they did just that. *Two bacteria kissed each other.* Right there on the screen, and there were kids in that theater too. And the next thing you know, the bacteria turned into men and women. I don't recall if they showed the men and women kissing as that wouldn't have hit me as odd like the kissing bacteria did; somehow the kissing bacteria were more memorable.

So that was sex evolution. Kissing bacteria turning into men and women. If not for those adventurous bacteria, just look at all the fun we'd miss! No doubt those bacteria somehow knew they had an exciting future -- if they'd just evolve.

Fast forward about 20 years to another museum (or science church, if you will). This time my wife and I are in London, at the Museum of Natural History. And now we're talking about flying, because there's a big display on how flight evolved in birds. A big, dumb display. Because it mainly talks about wings, and there's a whole lot more to flying than just growing a pair of wings, but the Chuckie Dee Evolution Club doesn't want you to think about that. Just look at the display and let them do the thinking for you. But I'm afraid that's not me. So I started *thinking* about what I was seeing.

Now, we had just spent all this money to fly across the ocean in a big giant hunk of roaring metal called a jet, and here I am watching a video showing how wings evolved, and thinking, "Man, Mother Nature could have saved us all that time and trouble having to come up with jets if we just could have done what these guys did!"

"These guys" were lizards. And they learned to fly and became birds. I mean, it was RIGHT THERE on a science video they were showing as part of the display! And the makers of the video presented two stories about the evolution of flight, because they weren't quite sure which story was right, so they made up two stories, and let the watcher decide. I liked that. I could decide how birds became birds and learned to fly. I didn't need anyone to tell me one way or another. So, what were the choices?

Well, on the one hand you had lizards *running downhill*. Once they got to cruising speed, they started flapping their arms. And...the video actually SHOWED their arms turning into wings, and they took off into the wild blue yonder! So, if it was on a "science" video, it MUST be true, right?

On the other hand, there were the tree lizards. Now, these guys learned to fly by *jumping* off branches! When they realized they were falling, they started flapping their arms, and sure enough the video actually SHOWED their arms turning into wings, before they hit the ground, in which case they would not have been much use to Evolution, but the video makers somehow missed that point. This was a "science" video, after all, and somebody must have known what they were talking about, or they wouldn't have made it, and it certainly would not have been up to snuff for a prestigious institution such as the London Museum of Natural History, right?

[Break for tea.]

Ok, we're back. So, what's my point in relating these stories about sex and flight? THEY'RE STUPID. REALLY stupid. But, only REALLY SMART PEOPLE believe them, and they belong to the Chuck Darwin Evolution Club. And once you're a member of the CDEC, it's very hard to get out. They have ways of keeping you in. I know. I used to be a member. And CDEC members want YOU to think that THEY are really smart, so that you'll just accept what they tell you without your having to think about it. They'll tell you how smart they are all the time, in fact. Recently I watched a video presented by a well-known atheist speaker and writer, and before he could even open his mouth he had already received praise to high heaven (hmm, not sure where that would be for an atheist, but carry on...) from the moderator about his wonderful academic achievements and what a smart guy the speaker was because he knew a little more than the rest of us about all there is to know.

The speaker then started spouting about what a wonderful "story" evolution is. At least he was smart enough to know THAT! Then he told us all about how bacteria turned into people. Yeah, that's right. He said that bacteria were our ancestors. And instead of being *laughed* at, the brainwashed audience *clapped* for him. Ask yourself why. Are we really all that gullible? Are we confirming that there really IS a "sucker born every minute?"

Let me give you just one reason why stories like how lizards learned to fly are stupid. As I said above, flight is more than just wings. It's like saying if you just put wings on a big, wide, long tube, it'll be able to fly people across the world. Birds need more than wings to fly. They need just the right feathers, the right bones, the right muscles, the right respiratory system, the right weight, the right shape, and - get this - the ABILITY to fly. Or did they just evolve that from thin air (a pun, yes, a bad pun)? They need their brains to be wired up for flight. So, where did THAT come from? An explosion? THINK about this stuff. Don't just accept it because an "expert" told you so.

Let's talk about birds some more. Birds build nests. Any Chuckie Deeists ever think about where birds got the *ability* to build nests? Did it just pop into their brains, and they just "happened" to be able to do it, just like that, or did the idea evolve into their brains gradually – stick by stick or straw by straw? And where did they live *before* they had nests? And what about every other creature that builds nests, or other habitats, like bees (I had to get them in there somewhere to complete the chapter heading)? Did the idea of how to build nests just pop into their evolved minds one day? EVERY ONE OF THEM EVOLVED THE KNOWLEDGE OF HOW TO BUILD THEIR NESTS??? *AND JUST WHERE DID THEY LIVE BEFORE THAT,* pray tell? How come we don't have millions of fossils of attempts to build nests as creatures evolved this knowledge? Because it's STUPID, that's why. It's dumb to think that that ability just popped out of nowhere, or slowly evolved into the minds of animals and bugs and other creatures while they were trying to "survive."

CHAPTER III

The Only Thing Evolving Is the Story Itself

As I said earlier, it seems like every other month a new "pre-human" surfaces, or some other new fossil or astronomical fact is discovered, and they add or subtract a million or a billion years to the age of the universe and mankind. Nowhere else can you throw numbers around like evolutionists do. Wouldn't it be nice if your banker could add a million or two now and then at whim? How about the IRS throwing an extra few hundred thousand in your tax refund? But when it comes to Evolution, after all, who's counting and who really cares about the millions or billions? They tell us that we're all stardust and part of the Universe, and how these wonderful evolutionary discoveries are going to help us know more about "who we are."

But let me ask you: If all we are is organized stardust, WHO CARES? And WHO CARES what Mr. PhD with ten other degrees has to say about it? Why should we even listen to what they have to say if we're all just meaningless specks of dust on a speck of sand floating in the middle of an ocean of nowhere? No, instead we've made those (like Carl Sagan, for instance) who tell us (with great authority) that we're nothings floating in an ocean of nothingness into our modern-day gods. Sagan said, **"The Cosmos is all that is or ever was or ever will be."** Anybody ever ask him, "Umm, excuse me sir, but just how do you know that?" No, because he's an expert on what he doesn't know, and who are we to question him?

Just more bull twaddle to shovel on the heap. Right, now tell me how knowing the universe is 12 billion years old (or whatever the current popular age is – you can find estimates between about 8 and 20 billion years and it's still changing) helps me know who I am? Is that what you call New Age Old Age philosophy? How does the myth that the dinosaurs were killed off by an asteroid help define why I'm here? How do a couple of ape bones found in Africa tell me something about my future? How does some prediction of how the universe will end in 15 billion years help give me strength to face each day? Not to mention the question of exactly how some puny human knows what's going to happen in 15 billion years. If we knew what was going to happen TOMORROW, now THAT might be helpful, as any astrologer knows (my next book just might be *Astrology Is Stupid*, but let's move on here; we're already stepping on enough evolved toes and astrology is big business just like Evolution). Maybe these billions of years guys are missing the boat and could be making a few bucks telling us about *tomorrow* instead of billions of years from now, *n'est-ce pas*?

(Oh, speaking of big business, the moderator in the atheist-glorifying evolution video I mentioned earlier did a five-minute pitch for his organization, asking for support and subscriptions to their magazine and plugging the benefits of joining them, then in the next breath criticized churches for competing to get your dollar. I like to show evolutionists who mock religious organizations for asking for money where the atheist- and evolution-focused organizations do the same (check out their websites!), and that usually shuts them up about that issue.)

Finally, there are lots of good, serious scientific arguments against evolution. He who seeks finds. While I'm pretty well versed in all that and have read and studied dozens of books and articles and watched numerous presentations (live and otherwise) on the subject from all sides of the issue, along with teaching, speaking and writing my own articles about it, my point here is nothing more than to demonstrate that Evolution itself is a silly idea and religious in nature, though perhaps that was not as obvious at one time. But we're beyond that now. Lots of ideas have come and gone in science, and this should be one more that needs to GO.

CHAPTER IV

Chuck Darwin Evolution Club Tactics 101

Or, How to Intimidate Those Who Don't Believe Your
Mythology

Now, I know exactly what's going on with some of you right this moment. I will often be referring to Evolution as a religious myth, which is exactly what it is. And many of its adherents – meaning some of YOU - are the worst fanatics around. How do I know? From over 35 years of confronting you, for starters. What you'll do is come up with all sorts of tricks to try to divert the issue from the stupidity, irrationality, and illogic of Evolution theory on to other things, and try to make ME look dumb and uneducated. You'll try to smear my character, or challenge my "qualifications." You'll accuse me of being this or that. Or just curse me, as often happens. I guess that makes you feel more intellectual.

Here are some of the classic diversionary tactics (call it going off on rabbit trails, hand-waving, smoke and mirrors, whatever) that you Evodelusionists like to use:

EvoTactic #1: "You're so dumb you don't know the difference between Darwinian evolution and cosmic evolution. They're NOT the same, you idiot!"

Oh yeah? SO WHAT? Now *you* don't have to explain **how nothing became something and turned itself into everything**? Yes, you still have to do that. Most evolutionists nowadays see a progression from cosmic evolution to the evolution of life because they've been backed into a corner and know they just can't start off with ready-made life, but have to concoct some fairy tale about where that life and the material that makes it up came from in the beginning. Hence the "we're all stardust" baloney. If you're a Darwinist you believe cosmic evolution. If you're a cosmic evolutionist, you believe Darwinism. So I've lumped cosmic and Darwinian evolution together under the heading "The Mythology of Evolution." Get over it and let's get back to work on **how nothing became something and turned itself into everything.**

EvoTactic #2: "So you are denying that things change? Are you serious? Evolution is all about 'change!' Certainly you wouldn't *deny* that things CHANGE, now, would you?"

I love that one because it's a classic example of how the *words* of evolution have evolved over the years (let alone the story line). It used to be that "evolution" signified the upward climb in complexity from single-celled organisms to man (on the lines of the ancient Greek "Chain of Being.") Since evolutionists can't demonstrate that that really happened, they decided to re-define evolution as "change" because *who can deny that things change?* Ok, I'll admit things change. What I want you to show me is how particles turned into plants, animals, and people all by themselves. If you can show me how THAT "change" took place, we can talk. I want you to show me **how nothing became something and turned itself into everything.** Got that yet?

EvoTactic #3: "Oh, he must be one of them Creeayshunists!" (translation: Creationists).

The Evodelusionist reasoning is this: "If I can just demonstrate that he's one of them Creeayshunists, that gets me off the hook and proves that [MY religion of] Evolution is true." So all you have to do is pin a label on anyone who disagrees with your precious myth, and somehow that vindicates and validates the myth, right? Wrong. You still have to show me **how nothing became something and turned itself into everything**.

Pinning a label on me may make you feel better about yourself and make you think that you're freed from having to give an answer, but that tactic's not going to work here. We're focusing on Evolution, not creation, not God, not religion. Except for the religion of Evolution. From nothing to everything all by itself. I want to know how it happened. I don't want promises that eventually "science" will find out. You can promise me that forever. I want to know NOW how that happened, so tell me.

At this point I should direct a word toward those who promote so-called "theistic evolution," which is the idea that God used evolution to "create" everything. That is the ULTIMATE cop-out, and nothing more than an admission that you recognize the Evolution myth is impossible without divine or some other (intelligent) intervention (same as those who promote "panspermia" or the idea that life evolved elsewhere then "seeded" the Earth, which just begs the question). An Evolutionist by any other name is still an Evolutionist, and most Evolutionists want nothing to do with ANY god or intelligence jumping into their myth. Nice try, though.

EvoTactic #4: "You people are ANTI-SCIENCE!"

Uh oh! What could be more devastating than being called ANTI-SCIENCE?? That'll show *you* a thing or two! Of course, this is just one more attempt at insulting those who don't belong to the religion of Evolution to make the accuser feel more important and intelligent and part of the "in" crowd, and, like Evolution itself, it has no basis in fact. I myself have always had a love for science, particularly nature studies, astronomy and geology. That's what got me into, and out of, the Evolution religion to begin with. Evolution ISN'T science. It's a belief system. A wrong, false one. Anyone who is REALLY interested in following the scientific method, and not their personal biases or those of the Evodelusionist Elite, should dump the theory without hesitation.

EvoTactic #5: On those same lines are those who will try to throw up another smoke and mirrors blockade by bringing in the Bible and saying something like, "You believe in a book that's 2000 years old and written by men?"

Despite the fact that no one ever said a word about ANY book other than Darwin's, which of course was written by one fallible, angry man, that's yet another attempt to get off the subject of whether Evolution is true or just stupid. On my website, EvolutionIsStupid.com, which was the precursor of this book, I did *everything possible* to keep the topic focused on Evolution and OFF any discussion of the Bible, or Creationism, or my personal beliefs, or religion itself (other than the religion of Evolution) for that matter. But the Evolution fanatics can't help bringing up those subjects because they think that automatically excuses their belief in their own stupid mythology. So I would ask them, "Why should it matter how old any book is, if it's true? I mean, there are lots of old books that we still accept as factual, no?" Then I ask if they believe in a book that's around 150 years old (Darwin's), and whose "science" has been demonstrated to be patently false? Never get an answer to that one for some reason.

EvoTactic #6: "What are your credentials? Are you a REAL scientist?"

You see, only a REAL fireman could know that it's stupid to play with fire. And only a REAL doctor could know that it's stupid to drink poison. So if you're not a REAL scientist, you can't possibly know Evolution from matter to monkeys to man is a stupid idea with no foundation in reality. Even if you have a dozen academic degrees after your name, if you don't believe the Evolution myth, then you can't possibly know what you're talking about or be a REAL scientist. After all, we Evolutionists worship REAL scientists and we know they have all the answers, and even if they don't we have FAITH that one day they will. That's our FAITH in action (though we'll deny to the death that "faith" has anything to do with it).

EvoTactic #7: Which brings us to another one of my favorites: "This argument pits FAITH against REASON. We science people look at the FACTS. And there are MOUNTAINS of evidence for evolution!"

When I confront those who use that latter assertion with a request to provide just a few rocks from the "mountains," the Evolutionists are hard pressed to come up with even a few good grains of sand. They've been using the same pathetic, generic arguments for years now, like "Look at the fossil record! What about mutations? And speciation! And…and…and CHANGE!!!"

Oh, and that "Reason" word will then enter the discussion. That's because, according to Evolution religion fanatics, it's perfectly REASONable to believe that **nothing turned into everything and made itself all by itself**, and here *we* are to contemplate it! That's REASON at work, isn't it? There's no *faith* involved in THAT belief, now is there? After all, there were lots of scientist-gods and other observers around to record it all, right? No, it's not right.

(By the way, take any history book that starts with the evolution of the universe and of humans. What you'll almost invariably find is a few pages dedicated to the "billions of years" that the universe and man were "evolving," (usually with some cutesy diagram or other) and then the rest of the book, perhaps hundreds of pages, will be devoted to REAL history, from about 6000 years ago to the present. Tell you anything? It certainly did tell *me* something when I was questioning Evolution. But I guess if you want to believe that man had nothing to do for a couple of million years but run around chasing big animals with spears then trying to figure out how to cook them, go right ahead. Oh, I guess we were inventing the wheel too so that took a little time.)

EvoTactic #8: Here's another: "I guess you don't accept **plate tectonics** or **gravity** either, huh? Well, huh?"

Ok, let's go over the "reasoning" again here. See, if I can divert the issue off of particles popping into existence and turning themselves into plants, persons and everything in the universe all by themselves, that gets me off the hook. So, let's try to make THIS GUY look stupid by accusing him of not believing perfectly testable, repeatable scientific facts like the effects of gravity, thereby skirting the real issue, which is just **how did nothing become something and turn itself into everything**? We certainly can't give up our precious Evolutionary religious myth, which is not testable, repeatable science.

I experience gravity every day. Can't see it. Don't know what it is, but I know it's there by its effects. I don't experience evolution every day. Neither does anyone, or anything, else. The only effect I see around me is that everything deteriorates (a testable science concept called "entropy"). I experience that in my own body every single day – I ain't gettin' any prettier, that's for sure. No further need for proof than a mirror. I don't see myself, or anything else, evolving into something better or more complex all by itself.

EvoTactic #9: Then there's the classic "Don't bore me with the facts" attitude, like one feedback writer to whom I recommended a book BY an evolutionist who was critical of the theory of evolution. She Googled it, read all the (evolutionary fanatic) articles that criticized it (without reading those that supported it), "skimmed" the first 20 pages of the book and that was enough to satisfy her that her belief in the myth was valid.

EvoTactic #10: If all else fails, we resort to name calling.

I've been called just about everything in the book since I left the Chuckie Dee Club – idiot, moron, retard, stupid, fool, a—hole, and worse. That usually gives the name caller the satisfaction of knowing he or she is FAR superior to the non-Evolutionist. Of course, some of us have evolved thick skin to help us survive the attacks of the blood-seeking evozombies (uh oh, doing a bit of name-calling myself there, I guess!).

So, regardless of whether I'm a Creeayshunist, or a non-scientist, or one who doesn't believe in gravity, the question still remains:

How did nothing become something and turn itself into everything?

CHAPTER V

Mysteries of the Universe Revealed!

You are about to be introduced to mysteries of the Universe (another nutcase tactic that sells books), hidden until now within the confines of the Chuck Darwin Evolution Club. You will realize that you may well be part of the Club, but do not even know it. I have penetrated their Inner Sanctimoniums (yes, I know it's not the right word so sorry, you can't use that one to prove I'm "uneducated" – just try to calm down), and I have exposed their priests (not literally), learned their tricks, memorized their stories, and ESCAPED TO TELL ABOUT IT! We will follow a logical Evodelusionist progression from simple to complex, which is just what Chuckie Dee Clubbers say they do, hoping that their hearers won't actually **think** about what they're saying. Here we go...

Where Did We Come From?

Now, let's get serious here. Ok, let's not. I've read lots of material both for and against the theory of evolution. Most of the opposing material takes evolution way too seriously. I realize there's a need for that, but I think we need to point out the fact that there's plenty to laugh at too, and then laugh at it. Like bacteria kissing and wings popping up on falling lizards, for instance. But that's only the tip of the coprolite pile (if you don't know what coprolite is, you'll have to look it up).

Let's start with the Big Bang, because starting with already living things like Darwin did is cheating. Now guys (meaning the males of our species), we like blowing things up and things that go "bang!" - isn't that true? I know I did when I was a kid. My brother almost blew off some fingers with a firecracker. Now *that* was cool, right? Cherry bombs in cans. Playing war games. Even though the atomic bomb was a fearful thing, deep inside we thought it was cool. We used to talk about how Russia had enough A-bombs to blow up the world seven times over. Didn't matter whether or not it was true. WHOA! SEVEN TIMES?! Now, THAT'S POWER! We gotta catch up with them Russkies FAST. (This was back in the '60s when it was ok to not be politically correct so yes, they were Russkies.)

When we weren't blowing things to smithereens, we were figuring out how to control explosions. That's how cars were invented. And rockets. We know cars and rockets are cool, because they're run by controlled explosions.

So, what if someone came up with the Mother of All Explosions? Well, of course it took a Russian to do it (I'm talking Cold War era, now, so keep in mind we didn't have to be PC). Physicist George Gamow developed the concept (initiated earlier by Georges Lemaitre, a Roman Catholic priest), of a sort of explosion that made the universe, and an astronomer named Fred Hoyle thought it was a dumb idea and sarcastically called it the "Big Bang." Obviously Hoyle did not know *cool* when he heard it. I mean, an *explosion* that made everything?! Who could top THAT?

Ok, let's get something clear now. Some guys in the Chuck Darwin Evolution Club don't like anybody calling it an "explosion." They say the universe just started *expaaaanding*, like a balloon that's being blown up. Well, if you blow up a balloon enough, what happens? It EXPLODES. Why can't these guys see this? Oh well... (Yes, Chuckies, that was meant to be tongue-in-evolved cheek so again, just keep calm). And if that's not bad enough, they say the balloon started out just as a point, about the size of a period like the one at the end of this sentence. Are you following here? EVERYTHING - including you, me, cats, rain, planets, trains, the Rolling Stones, hamburgers with the works, EVERYTHING started out from something the size of this (.). If you are a member of the Chuck Darwin Evolution Club, and you do not believe this, you will face serious consequences.

And now, to reveal the hidden secrets of "The Club." But before we get going, I need to just nip this one in the bud: I'm fully aware that Darwin focused on what was already there, not where it all came from in the beginning (though he wondered about that just like the rest of us). However, as you will soon see, the fact that Darwinism begins with what's ALREADY THERE is…well, umm… a bit of a problem for Darwinists.

No modern Chuckie Dee Evodelusionist Club member would deny the fact that we need to go back beyond what's ALREADY THERE to *where* what's already there came from to begin with, so we're including all that in the overall evolutionary scenario. I've already gone over this once, but figured it would not hurt to repeat it. Do try to remember the phrase "IT WAS ALREADY THERE" for future reference, though, if you will.

Chuckie Dee Club Secret #1: DON'T ASK WHERE THE "PERIOD" CAME FROM.

You see, Chuck Darwin Evolution Club members don't like squirm-inducing questions like that. The period was ALREADY THERE, OK? Space supposedly did not exist yet, even though I don't understand how the period was not taking up space and neither do you. But there was just *nothing* before the period existed. And the nothing became the period, then the period started to expand, or be blown up, or whatever happened. And eventually hamburgers came out of it and don't you worry, we'll come up with a story about how that happened, so just be quiet or you'll have your membership revoked.

So that's how it all started. Period. But how did hamburgers come from it? Well, as we noted already, the period started to expand... Wait!

Chuckie Dee Club Secret #2: DON'T ASK WHY THE PERIOD STARTED EXPANDING.

It just did, ok? Nobody was there. Nobody saw it. We can't duplicate it. We believe it *by faith* and we can kind of prove it on a blackboard. It just happened. Believe it and shut up!

Stop asking questions and let's get back to the mysteries and secrets of the Universe. This is a really neat story that Evolutionists have made up. At least they think so. Ok, so the period started expanding, creating space as it went. And then matter came into existence... Wait!

Chuckie Dee Club Secret #3: WE INVENT UNPROVABLE, UNTESTABLE STORIES ABOUT HOW MATTER CAME FROM THE PERIOD, WHICH CAME FROM NOTHING, MEANING MATTER CAME FROM NOTHING, WHICH WE DENY BECAUSE THEN WE'D HAVE TO COME UP WITH WHAT WAS THERE BEFORE THE PERIOD, WHICH WAS NOTHING, SO WE JUST SAY THE PERIOD POPPED INTO EXISTENCE, AND WE DON'T WORRY ABOUT WHAT WAS THERE BEFORE THAT. GET IT? OH, AND WE ALSO CALL THE PERIOD A "SINGULARITY" BECAUSE THAT MAKES US SOUND SMART.

If you ask what was there before the "singularity," you'll just be told that it doesn't matter where matter came from because matter only matters because matter exists. Get it? You have just been introduced to an unwritten rule of the Club: Don't ask too many questions, and if you do, don't expect any logical answers. Just accept what you're told by the "experts" and move on. After all, they have worked all this out on blackboards, and it doesn't matter whether reality is involved or not. (Yes, blackboards. Scientists, like Einstein for instance, are almost always pictured in front of blackboards. Like doctors always have stethoscopes and football fans always have beer in their hands.)

Ok, so matter then started collapsing and spinning. Oh no, here we go again...

Chuckie Dee Club Secret #4: WE DON'T KNOW WHY IT ALL STARTED SPINNING. IT JUST DID. AND WE HAVE INVENTED NICE STORIES TO EXPLAIN IT.

Look around you. Everything spins. Particles spin, planets spin, moons spin, galaxies spin, and it appears the whole universe is spinning. Things just don't spin on their own, right? It's a scientific principle that angular motion (spinning), as opposed to straight-line motion, can only be produced by an outside force, right? But the Club says that's EXACTLY what everything did - it started spinning on its own, and you need not question why. Which is stupid, but you have to be really smart to believe this stuff. Just don't THINK about it, and you'll be fine. The universe spins. Within the universe, galaxies are spinning. Within those galaxies, stars are spinning. Around those stars (at least one for sure), planets are spinning. Around those planets, moons are spinning. Your head is now spinning. And it should be, because spinning does not start on its own, with no outside influence, so it's stupid to insist that it did.

Which brings to mind a very important break we need to take at this point, so we can discuss something up close and personal in the history of humankind. Just how did stupidity evolve, anyhow? I don't recall ever seeing an evolutionary explanation for that one. So we'd better invent something.

CHAPTER VI

Stupidity

You will notice that the word "STUPID" is being used frequently in this book. That is an offensive word, which is good. Now I'm a nice guy, and would not call anyone stupid, especially if he's bigger than me. But boy, do we believe some stupid stuff. And often we don't even realize how stupid it is till it's too late. Just look at some of the stuff of the past that was called "science." Ever heard of "phlogiston"? Not too long ago we laughed at the ideas of television and space travel. And how about Astrology, which has no basis in reality? Not only are we made of stardust, but some star a billion light years away is controlling your life right this moment. You believe that? Based on WHAT? They told you you'd find romance today, and some cute guy winked at you and suddenly the astrologers were right?

Science is always changing its tune too, and that's fine, because it's supposed to be about testing, and proving, and demonstrating repeatability or falsifiability. NONE of which is possible with the Big Bang and Evolution from mud to mom, but they just can't admit that and get over it. They have to protect the myth.

Now, we might believe something just because someone told us to, or we might believe it because we're really convinced it's true, or we might believe it out of fear (for example, of peer reprisal or that we'll lose our job or position, or whatever), or we might believe something just because we don't care and don't want to be bothered really thinking about it because we have "better things to do."

However, the questions that the religion of Evolution seeks to answer are the kind that we just can't ignore because they're questions we all ask, and for which we all seek answers: **Where did I come from? Why am I here? Where am I going?** According to the Evolution myth, you came from slime that turned itself into Slim, you're here to pass on your genes (even though we have no idea how they evolved), and you're going back to the slime you came from as fodder for future evolution. Now, AIN'T THAT GREAT? So why do we even BOTHER getting all uptight about this stuff then? Do CATS get all upset worrying about who they are, where they came from or where they're going after they die? There must be something different about people, that we're so interested in these questions, right? Could that be possible? Naw. We're just another animal. Sure, we build cities and fly rockets to the moon, but we're just another animal. Stupid, is what that is.

As I told you, I used to belong to the Chuckie Deeist Club back in the 1970s. I first began to unlock the mysteries of the Club before going to college, when I took a trip out to the western USA with two of my cousins. We all hopped into my uncle's enormous Olds '98 complete with cruise control which was cool even then, and pointed the car Westward. That is where I had one of my most memorable life experiences that changed me forever. Stupidity played some part in that experience, but wound up making me a little more able to handle what life had brought my way, and what life would bring my way, so in the end, I was a little smarter, anyway. Some of us learn. Some don't.

After a few stops to see various sights like Devils Tower in Wyoming, where I now live, my two cousins and I went to

Rocky Mountain National Park in Colorado. We had read in a tour guide about Long's Peak. At over 14,000 feet above sea level, you could climb to the top of the mountain without climbing gear, we were told. So that's what we wanted to do. None of us had any kind of gear. No backpack. No knives. No food. No water. I guess we had good climbing shoes, but that was about it. And we had a dose of dumbness, which would also help out. If we'd have known what we were facing, smartness would have told us not to do it. Oh, I did have a Kodak Instamatic Camera with me too.

We set out on a path marked for Long's Peak, but soon came to a Y in the road, and man, we were stuck. Although there was a sign there supposedly pointing the way to the peak, to me it said take one branch of the Y, but to my cousins it said the opposite. We had a low-level discussion about it for a few minutes, and decided the best thing for a stubborn dog like myself to do was go my way, and they'd go theirs. And that's what changed my life, and I've retraced those footsteps many times since that time, and will follow them to my grave, I'm sure. In fact, that's why Robert Frost's poem, "The Road Not Taken," became my favorite (I'll be mentioning that again shortly).

As I started up the path I was sure would lead me to triumph on the summit of Long's Peak, an older man was coming toward me. I have often looked back on that and wondered if he was some sort of angel. I just had a weird feeling about him. There was nobody else around, and we exchanged a few words. He looked at me, chuckled, and told me it was a difficult climb ahead, though he thought I could make it. I was young and invincible, and was not about to let that stop me, though the rocks and mountain before me looked pretty formidable.

Kids do stupid things (adults do too, but that's for another time). And they can get away with it sometimes, just because they're kids. So the next thing I know, I'm facing a ledge, and if I don't cross that ledge, I might as well turn back. I had to cross that ledge if I was going to make it to the top. But this wasn't just any ledge. It was about four inches wide, with about a forty foot or more sheer drop below it. Now, there's stupidity with a goal, and there's stupidity for the sake of stupidity. Mine was the former. I was going to cross that ledge, and nothing was going to stop me. My goal was to reach the top of that mountain, and show my cousins and the world that I had taken the correct path, and that I was right. As I often still tell people, "You don't have to agree with me - just admit I'm right!" (I'm thinking of writing a book by that title too).

Heart racing, I stuck my camera inside my belt, and placed my feet sideways on that ledge, pressed myself against the rock, and started inching my way across. Was I scared? Naw! Me? I was petrified! But I was not going to let this mountain defeat me. Then something frightful happened.

My camera slipped out of its nest, and I froze.

A few seconds later, way down below somewhere, I heard a "PLINK!" and knew that was the end of my camera and any memories recorded on it. But the really scary thing was the split-second decision I had to make to NOT go after the camera as it started to slip from my belt. In the short time between THINK and PLINK, I could literally have died trying to save the thing. Instead, it was now that much easier to slink the rest of the way across the ledge, and I had one less item to preoccupy me as I was trying to cross it.

Obviously I made it. That's why I can be writing about it now. You know, when you go through something like that, it's not pleasant, but when it's over, that's all you want to talk about. Just like a hurricane, or an earthquake, a battle, or some other life-threatening trial.

But the adventure was just beginning. When I got closer to what I thought was the top of the mountain, my eyes looked up, and my heart sank down. Because until that point I was sure that the snow bank ahead was the top, but the awful truth was that the top had been hidden from my view. Now I had to decide: Press on, or give up? I don't give up without good reason – that's just not me. Though I had no idea what time it was, had no water or food (I was eating dirty snow to quench my thirst), I was getting to the top of that mountain, and I was going to live to tell about it. At least I hoped so.

Finally, after a forever of climbing over rock, gravel, and snow, the summit was in view, and I knew I would make it. The beauty of the surrounding terrain, the panorama of snow-capped mountains all around, hit me so hard, I just got on my knees and talked to God, even though I didn't know who God was at that time. My heart was racing from the lack of oxygen at that altitude and strain of the climb, and I found some ice to eat beneath a rock to quench my thirst. And then I looked over, and saw that the mountain across the chasm before me seemed just a bit higher than the one I was on. And there were two specks moving up the sheer side of it! I knew they were not my cousins, unless they had gone out quickly and bought some climbing gear, but they were other humans, and I was glad to see them. I cupped my hands to my mouth, and yelled over, not knowing if they'd hear.

"Helloooo!!!!"
One responded. "Helloooo!"
"What time is it?!!!"
"Mid die!!!!!!"
"WHAT?!"

"MID DIE!!!"
"I DON'T UNDERSTAND YOU!!!!!"
With that, they yelled at me about an octave higher, with
ascending, then descending tone,
"NEWWWN!!!!"

I then understood that these guys must be Australians, and
they were saying, "Mid day!" and "Noon!" So I yelled my
"thanks" and was amazed that I had made it up my mountain
so quickly, as I thought it was much later in the day.

But now I had to get back down. That turned out to be more
difficult than the climb up, as I slid down mini avalanches of
gravel, trying to dodge huge boulders. There was a way to
avoid the ledge I'd had to traverse on the way up, and I found
it and took that route, and soon enough was back on level
ground, walking down the same path I'd started up early that
morning. And who should come walking toward me, but my
two cousins?! Amazingly, seven hours after we parted, we
met again at almost the same spot. We each gushed out our
experience, and as it turned out none of us had taken the right
path to Long's Peak, and they wound up going up another
mountain, just as I had done. I had climbed 13,911 foot Mount
Meeker. But from that experience I'd learned some lessons
that I would apply to the rest of my life.

The first lesson was that sometimes you just need to follow
your heart and not worry about the difficulties. I could have
joined my cousins. I'd have been safer. I'd have had the
companionship of other human beings. Even if I were sure
the "other route" was the right one, at least I wouldn't be
looked at as a rebel, or as wanting my own way, or as being
disagreeable, or as a fool for going off on my own, especially
with no water or supplies.

But I believed the route I took was the right one, and pursued it to the end (like boxer Rocky Balboa would say, "I went the distance"), and the experience I had along the way, while difficult and dangerous, doubt-filled and daunting, in the end proved to be the best route I could have taken. In fact, that's what prompted me to adopt Robert Frost's "The Road Not Taken" as my life poem:

> *Two roads diverged in a yellow wood,*
> *And sorry I could not travel both*
> *And be one traveler, long I stood*
> *And looked down one as far as I could*
> *To where it bent in the undergrowth.*
>
> *Then took the other, as just as fair,*
> *And having perhaps the better claim,*
> *Because it was grassy and wanted wear;*
> *Though as for that the passing there*
> *Had worn them really about the same.*
>
> *And both that morning equally lay*
> *In leaves no step had trodden black.*
> *Oh, I kept the first for another day!*
> *Yet knowing how way leads on to way,*
> *I doubted if I should ever come back.*
>
> *I shall be telling this with a sigh*
> *Somewhere ages and ages hence:*
> *Two roads diverged in a wood, and I--*
> *I took the one less traveled by,*
> *And that has made all the difference.*

What's the point? The point is that sometimes when we take the road less traveled, the one most people don't take, or won't take, or are too afraid to take, that's the road the makes the most difference in our own lives, and the lives of others around us. Think of all the people throughout history who didn't just sit back and ride with the tide, but swam against it and changed the world for better or worse!

I also learned about not giving up. I don't know how many times since that mountaintop experience I have thought my goal was near, only to see it off in the distance, knowing I still had plenty of climbing, bruises and work ahead. Life is like that. Get used to it. If you get knocked down, get up again and move on.

So if you want to go on forcing yourself and others to believe something you know is not true, if you're afraid to go against the tide and speak out against falsehood and stupidity because others will look down on you for standing up for truth and you might sustain a few bruises along the way, **how can you live with that**? How can you deliberately continue to lead your and our children and others down a path that you KNOW is the wrong one, continually inventing stories just to save the party line? You've GIVEN UP. You've surrendered your will to the Evodelusion Club, and you refuse to admit it's so and do something about it. **I have never seen such examples of denial as evolutionists who look at the world around them and insist it's all the result of some lucky accidents, when they KNOW in their hearts that that's a LIE.**

CHAPTER VII

Spinmeisters Extraordinaire

We're getting too serious now. Let's get back to stupidity. Evolution is stupid and I am not going to pussyfoot and pretend it's not, so people will think I'm NOT stupid. I have over 35 years' experience with studying the issues involved in evolution from all sides, and the more I look into it, the stupider it gets. I once watched a video where Evolutionist priest Richard Dawkins was the main speaker. A participant in the audience asked him how the sexes evolved. He gave about a five-minute discourse saying, in effect, "I have no idea," but of course he couldn't bring himself to say that. Instead he made up stories of what "might" or "could" have brought about the sexes. Science? Not quite. Storytelling? Yep.

Now let's continue with the Chuckie Dee Evolution Club story of how we got here…

Ok, so everything was spinning. We got that far. But remember those explosions, guys? We're not done yet! Plenty more coming! The next explosions were stars. Now, keep in mind that the Big Burp was the Mother of All Explosions. But when stars explode, that's not exactly kids' stuff either! Here's what the Evodelusion Club says:

Inside the stars some elements were forming. Elementary, Watson, right? You bet, Sherlock! And then some stars started exploding way before their time, and spewing those elements all over the universe. Stuff like iron and other metals, which we would eventually need to make cars and other things that work by controlled explosions (not to mention grills to cook those burgers!), was spread all over the place. But then it started getting back together. Try to follow this, now. It's all been "proven" on the blackboard.

Now, if everything was flying apart, then why did it start getting together again, and why didn't it ALL come together, instead of getting together in groups that would become galaxies, other stars and planets? This is one of the myriad mysteries the Evolutionist has to invent "just-so" stories to explain. Oh, they'll always have some dumb story, and it won't make sense to you because in reality it only exists on paper with no connection to reality, and can only be understood by an elite priesthood, which probably does not include peons like you and me.

Ooooh, I can just hear Chuckie Dee Club members bristling at that one. They're gonna show ME a thing or two. No they won't. Don't worry about it. Evodelusion Club members get very angry when backed into a corner, but their bark is much louder than their bite. Evolution from particles to planets, plants and people, exists on paper and blackboards, and in the minds of the Evodelusionists. It does not exist in the real world. There is not a shred of evidence for it, **and it is totally unnecessary to the progress of science or the progress of anything else for that matter.** Science was progressing just fine without Evolution mythology, and will continue to do so when it's gone. Actually, given the level of scientific fluency of many of today's students, it's more than fair to say that since Evolution took hold, science literacy has made BACKWARD strides. (Ironically (or maybe not), as I was doing the final editing of this section, an article appeared on the Internet concerning proof that humans really are getting dumber!)

Back to explosions... Yes, everything started getting together again, and condensing into more stars and even planets. Easy as that! You may have seen a science film or some video that showed it. Things always work out just right in science films, and we all know that whatever we see in a science film MUST be true, right, 'cause the guy in the white smock in front of the blackboard said so. You'll see (in the film, that is) these nice clouds of dust and debris (elements) that came from "somewhere" (don't ask). They start getting smaller (condensing) all on their own, for no good reason (any dummy knows that if a gas is left to its own it does not come together and condense into a solid ball, but that doesn't matter to the Evolution Club). Then the balls form into stars and planets, just like ours! How convenient that they do that!

I like planets. They really blow evolutionists out of the water (so to speak). At least the ones we know do. If what Evodelusion Club members say were true, all the planets of our solar system, along with the sun and moon and other planetary satellites, really should be made of pretty much the same stuff, spin in basically the same direction, be angled toward the sun the same way (most logically they'd have an axis parallel to the sun's), and have pretty much circular orbits. But that's not the way it is. Some spin backward (compared to the others), others have various axis tilts, some are gaseous, some solid. Earth isn't like ANY of them except maybe Mars at best, and so on. It's like someone's out there trying to say, "HELLOOOO!!!"

But Evo Club members just can't give it up so here we go with more invented stories, like that comets somehow supplied most of the water on Earth (kinda sorta missed the other planets, apparently), or that some monstrous piece of space debris (evidence of which we can no longer find, of course) knocked into the Earth and the moon came off as a chunk.

Now, the moon SURE LOOKS LIKE a hunk of Earth that was knocked off, doesn't it? No? Well, then, let's come up with a new "theory" like that a piece of the Earth spun off as the Earth was forming. Well, now, if that happened, wouldn't that piece just fly into space instead of ending up in a nice, nearly circular orbit around the Earth? Ahhh, but it happened BILLIONS of years ago, and so, once again, we can't prove that the story is dead wrong, because, CONVENIENTLY, it happened SOOOOO LOOOONG AGO!

Let's focus on the Earth for a moment. How much different from the rest of the planets can you get? You know, there are some people who believe that there was once a great Flood on the earth that resulted in all the strata and fossils. But not Chuckie Dee Evolution Club members. They can't see how that could have happened. But guess what? They DO KNOW THERE WAS A FLOOD ON MARS! Even though there's hardly any visible water there! Isn't that amazing? So people who believe in a worldwide Flood on an Earth that's almost three-fourths covered with water are dummies, but people who believe in a flood on Mars are intelligent!!!

Oh, and how convenient that the tilt of the Earth's axis is just right, so that the surface area of the landforms and water in the northern and southern hemispheres "just happen" to allow for moderate temperatures so we don't fry in the summer and freeze in the winter. Cosmic evolution is just so... so... THOUGHTFUL.
****

Speaking of intelligence or the lack thereof... Let's move on from stupidity for a bit and see if we can find some intelligence. Let's have a look at...

CHAPTER VIII

SETI, or Signs of Evodelusionist Temporary Insanity

All this talk of planets and smart people brings something else to mind: The Search for Extraterrestrial Intelligence, or SETI. Maybe you've heard of it. Scientists all over the world are waiting for a signal from space, something like this one:

OERONEONVEONVAISDFSBBTYOUDUMMIESBEIOHREN WAOHPROEBQT

If you analyze the above gibberish closely, you will see that embedded in it is a secret message from an unknown intelligence, namely: YOUDUMMIES. These guys are actually getting paid to listen for a signal from an alien being or civilization, and they are making fools of us earthlings. How?

Well, apparently it has not occurred to them that, if there really IS an alien civilization out there, they'll probably want to DISSECT US in ET science class, or EAT US, or just BLOW US UP for fun. Kinda like we do to each other. Oh, no, WAIT! Of course they'll be a NICE, ADVANCED alien civilization which has figured out how NOT to do those nasty things, and their signal will arrive on Earth at SETI headquarters just in time to save us from annihilating ourselves. A Savior from the planet Zork! GLORY BE!

What's *really* the point of what these SETI people are doing? They insist that there's no evidence of intelligent design in life on Earth, but they're looking for EVIDENCE OF

INTELLIGENT DESIGN SOMEWHERE ELSE in the universe!
Because without design, you have no messenger and no
message. They know that, but because they belong to the
Chuck Darwin Evolution Club, they cannot bring themselves
to admit that it is STUPID to say that you yourself show no
evidence of design, while looking for a message from a
civilization that does. But, let's get more practical and down-
to-earth for a minute...

I lived in Italy for a number of years. I used to love to listen to
peoples' reactions when they saw a sculpture or painting or
building that was beautiful and exhibited incredible human
talent. Take the David statue in Florence, sculpted by
Michelangelo. People would gawk at it, gush about what a
great sculptor Mike was, gloat over his talents, and just
generally oooh and ahhh with mouths wide open as they
walked around looking at this statue of a naked guy. But if
you placed a real, live, in-the-flesh human being in front of
them (hopefully sporting a loin cloth), of which a statue is a
poor copy at best, well, now, to most observers who would
have been indoctrinated and brainwashed by Chuckie Deeist
materialism, that human being would just be the product of a
bunch of accidents ("lucky" accidents, as Evolution priest
Stephen J. Gould would have said).

Is that STUPID? You decide. If I repeatedly throw a hammer
and chisel at a piece of marble... Oh, never mind!

Let's take a closer look at this a minute. No modern observer
saw Michelangelo create the statue, but the same observer
would have no problem believing he did it. The statue is just
a piece of stone which is formed into a COPY of the
EXTERNAL FEATURES of a human being, and has no brain,
no blood, no heart, no feelings, though it was MADE by real
human hands, directed by a real human brain using tools that
a real human brain devised. So to say the STATUE was

created is no problem, but if you say the hands, brain, and man that CREATED the statue were created, you are chucked out of the Chuck Darwin Club, banished forever, and you are branded as STUPID and UNEDUCATED.

One day in Italy I was talking with a guy in a wheelchair. He said he did not believe in a creator, and had no problem believing that evolution made everything. I pointed to a monument and asked him if he believed it had a creator. "Of course!" I pointed to myself and asked him if he believed I had a creator. No, he didn't. I don't care if this does not sound kind, but that is just plain stupid. Evolutionists KNOW it's just plain stupid, but they just can't bring themselves to admit it.

***

We need to get back to explosions. Ok, so the stars exploded, spread stuff (elements) all over, and some of it fell to Earth in very convenient quantities. Now, it's a good thing that happened, or we would not have hamburgers, let alone ketchup or relish. Or fries. Because that stuff started getting together to make more complicated stuff, and we come to…

Chuckie Dee Club Secret #5: THE WHOLE EVOLUTION OF LIFE STORY IS ILLOGICAL, AND WE KNOW IT'S ILLOGICAL, BUT WE CHOOSE TO BELIEVE IT ANYHOW BECAUSE WE'RE EITHER TOO PROUD OR AFRAID OF THE ALTERNATIVES.

CHAPTER IX

Come Together, Right Now, into Me

There is no evolutionist out there who could seriously believe the scenario I'm about to paint, but they ALL say they believe it, because if they did not, they'd be kicked out of the Club on an evolved foot in no time flat.

First, elements and compounds, including, very conveniently, water, fall out of the sky onto the Earth, which is cooling down from being mighty hot, and they all mix together in the big ocean. Now, any kid can see that if you put stuff like sugar, or dirt, or whatever, into a swimming pool, it doesn't come together, but rather SPREADS APART. This is especially convenient when kids do things like pee pee in a swimming pool (adults, of course, would never do such a thing). By doing that they are adding READY MADE ORGANIC COMPOUNDS like urea to the water; you don't even have to wait millions of years for them to form! And do they all get together and turn into cells? Uh, well... not really. Of course, you would not want a blob of urine following you around, turning into cells, and neither would your swim mates. But you see, if you believe Evolution, every rule can be broken, and even if you have an OCEAN, and not just a mere swimming pool, when Evolution comes into play, organic compounds GET TOGETHER in the water, instead of spreading apart like normal, everyday experience would dictate, therefore we have...

Chuckie Dee Club Secret #6: YOU ARE NOT TO QUESTION BROKEN RULES, ESPECIALLY IF THEY'RE SCIENTIFIC RULES.

So the rule that's broken (but don't ask) is that, instead of the stuff spreading out in the oceans, it CAME TOGETHER millions of years ago! And not only that, LIGHTNING showed up and started flashing all over, and when it zapped some of this stuff, it formed the (drumroll) **BUILDING BLOCKS OF LIFE**!!! Whoa baby! And I mean BABY. Think of a baby playing with building blocks. If the blocks just sit there, does anything happen to them? Umm, no. But if a child picks them up and places them on top of one another, something happens to them, right? Or if he or she makes a building with them, something has happened. Or, if they have letters on them, the child might form a word. Otherwise they just sit there. Right?

But **NOT EVOLUTION BUILDING BLOCKS**!! NOOO SIR!!!

Evolution building blocks BUILD THEMSELVES!!! That's why you have to watch out for lightning in that swimming pool, because if it hits those organic compounds, you don't know WHAT ooey gooey stuff might result! (Now let's not even consider the fact that no one has ever actually seen lightning CREATE anything, though there's plenty of evidence that it DESTROYS things, but that's another issue. Don't ask.)

Evolution building blocks require no talent, and need no intelligence to form complex things. They just started to get together and make everything we know, and who cares HOW they did that. They just DID! Despite the fact that we see no such thing happening now, nor ever did, Chuck Darwin Evolution Club members, with their secret knowledge, KNOW that the "building blocks of life" got together and built

life! They don't know what life *IS*, but they're pretty sure they
know the story of how it got here. Secret information,
available only to the anointed priesthood. The rest of us are
left to just wonder, and accept, what's fed to us by the
members of the Club priesthood.

And priesthood it is! Try this: Tell a Chuckie Deeist Club
member that he belongs to a religious system. Tell him its
priests are evolutionary scientists, its pope is Chuck Himself,
its churches are museums, and especially natural history
museums. The CDEC has beliefs about where we came from,
who we are, and where we're going – just like any religion.
Only in this case we're a bunch of "lucky" chemicals that got
together on their own and will return to the dust to fuel future
evolution.

Which brings to mind my favorite cowboy poem,
"Reincarnation," (*Cowboy Curmudgeon and Other Poems*, by
Wallace McRae. Gibbs-Smith 1992, p. 49). The poem traces
the journey of a cowboy's friend, Slim, who dies, is buried,
and shows up as a flower above the gravesite, which is then
eaten by a passing horse, digested, and eliminated by the
horse. In the end, seeing the digestive end-result of the
flower's journey laying on the ground, the cowboy says to his
'reincarnated' friend: "Slim, you ain't changed all that much!"

And that, my Evolutionary friends, pretty much sums up your
journey of existence. So, why bother getting your underwear
all tied in a knot about God, or creationists, or theology or
whatever, if that's all there is? (Unless maybe, hmm, maybe
you're hoping that's NOT all there is?) And why exalt and
applaud the high and mighty scientists who are telling you
(and going to extremes to PROVE to you) that you're a
nothing in the middle of nowhere going no place in
particular?

Ever asked yourself how they know that? Ever asked *them* how they know that?

And if it's all true, then WHO CARES??!!

CHAPTER X

The Faith of Billions Can't Be Wrong!

Let's go back a billion years. How does any evolutionist really know what things were like then? None of them was there to see it, but have you ever wondered how they can tell you all about it? That's like someone who didn't go to a ball game giving you a play-by-play recap. Would you believe them over somebody who was there? No? Then WHY do you believe Chuck Darwin Evolution Club members who weren't there but think they can tell you all about what happened millions and billions of years ago? I know why you believe them. It's because they're EXPERTS in what they *never saw and can never test nor repeat.* If you question their religious faith statements you are considered just an ordinary peon who needs to be educated in their religion, and you are made to feel like a fool if you don't join the Club and believe what they believe BY FAITH.

Now, you may have heard about a guy named Stanley Miller. He is one of the priests of the religion of Evolution, because he mixed some chemicals together in a neat-looking glass flask, zapped them with electricity for a couple of weeks, and collected the results, among which were (drumroll) **THE BUILDING BLOCKS OF LIFE**!!! Or at least, that's the story Chuck Club members want you to believe.

In reality, an experiment has to have a DESIGNER and CREATOR. An experiment does not make itself out of nothing, but you're not encouraged to think about that, and those two words are not allowed to be uttered in the Club. It doesn't matter if it's true or not. Clubbers just don't use those words. Now, if the designer/creator of that experiment (Stanley Miller) had not turned off the electricity at the right time (oh, and we're not talking lightning here), the "building blocks" would eventually have been destroyed, but you're not told that either. And the "building blocks" were nothing more than a few complex chemical compounds; had they been left in the flask, they'd still be sitting around somewhere doing nothing, which is exactly what they'd do forever if there were no builder to do anything with them. But you're not encouraged to think about that either. Just accept the party line and don't worry, be happy.

But the good Dr. Miller is one of the heroes and gods of the Chuckie Dee Club, because he made these "building blocks," which proves nothing to any thinking human being, but if you believe in STUPID evolution, you have to get all excited and say what a great thing Miller (and his partner Urey) did, otherwise Club members will scowl at you and ostracize you, because Miller's experiment is one of their icons, and you have to reverence it. Funny that no one has improved on it in over half a century, but let's not go there. And let's not even bring up the question of what would have happened to those "amino acid" compounds if they were dumped in the ocean, where life supposedly began.

Which brings us back to the ocean with the stuff ("building blocks") in it.

Now, the Club says stuff got together in the ocean or pools of water and formed what are called "protocells." Those are cells that came before real cells. Well, stuff did no such thing, and they know it did no such thing, but that's what they teach, that's what we learn, and that's what we must believe to be considered smart in our society, even though it's stupid, because the protocells would have just decayed and wasted away, and that's a demonstrable scientific fact/law, but keep in mind that breaking scientific laws is a hallmark of Evolutionary beliefs and fully permissible.

Ok, then these protocells got together in the big ocean, you see (but don't ask why or how because the best you'll get is an invented story), and they formed multicellular critters. Sort of like the dust and gas of space got together and formed stars. Are you with me here? The dust and gas exploded, then got together and formed planets, one of which was Earth, on which loose elements got together and formed cells. It was all VERY SIMPLE. We see it happen in science films all the time, and science films don't tell lies. (By the way, we'll get back to the kissing and sex eventually; so just be patient.) This is all so stupid to believe in that we should be completely embarrassed by it, but not only are we *not* embarrassed by it – we actually PREACH this stuff and smugly consider ourselves to be intellectuals if we believe it.

If all that is not bad enough, Evodelusionists are more than happy to IGNORE all the LIES that have been perpetrated in the name of their religion over the years. We have the FRAUD of the peppered moths. We have the FRAUD of Haeckel's embryos (say "ontogeny recapitulates phylogeny" and you'll *really* impress your friends!). We have the FRAUD of Piltdown man (based on a forged skull) and Nebraska man (based on a PIG's tooth). We have Neanderthal man formerly pictured as a brute, but now we're told that if we saw him on the street with a cell phone we wouldn't be able to distinguish him from any other dirtball with a phone. Whale evolution fraud. Made in China "feathered dinosaur" evolution fraud. Shall I keep going? But does any of this count in our Chuckie Deeist's minds? OF COURSE NOT. Want to hear the excuse? Here it is: "That's what science is all about!" Since the frauds, some of which were believed for YEARS, were eventually proven false, we're ok with that because "that's what science is all about." Got that? You tell the public a lie. Another scientist demonstrates that the lie is a lie. That's how REAL science works, so hurray for science! Vindicated again. And let's ignore the philosophy that ALLOWED us to believe those lies in the first place!

CHAPTER XI

Don't You Love Spontaneity?

Chuckie Dee Club Secret #7: CHANGE WORD MEANINGS SO AS TO CONFUSE UNWITTING MEMBERS AND OUTSIDERS.

Maybe you've heard of the "spontaneous generation" of life, and how people once believed that living things arose from non-living things, because, for instance, they noticed that maggots appeared in decaying meat, so they figured they came directly from the meat. Then finally Louis Pasteur showed that life could only come from life, which he demonstrated by enclosing meat in one jar, where flies could not enter, and leaving another jar with meat in it open, where flies could lay their eggs in the meat. He showed that you must already have life to beget life, or in other words, life CANNOT come from non-living matter on its own – that is, spontaneously generate itself from non-living matter.

But Chuck Darwin Evolution Club members have brought science back a couple of hundred years (and broken another scientific law), because they believe that life DID arise from non-living matter, despite the fact it was disproven long ago. They don't tell you that, however. Instead, they change the words "spontaneous generation" into a new word: "ABIOGENESIS," which is one of the many code words of the Club, wherein words are adapted (evolve, you know) to suit their agenda. For instance, as I alluded to previously, "evolution" used to mean that "simple" things became more complex with time. But now that Club members know full well that no such thing happens, they've changed "evolution" to mean simply, well, "CHANGE!"

Chuckie Dee Club Secret #8: THE ONLY THING REALLY EVOLVING IS THE THEORY OF EVOLUTION ITSELF.

We've already discussed this some in an earlier chapter. By that I mean that the only thing that's really changing and adapting is the concept of evolution. Evolution is like clay: you can mold it, adapt it, and fit it wherever and however necessary, so that the theory adjusts to whatever twists and turns are necessary to its form, function, and propagation.

Let me take you back again to my college days in the early 1970's, and what got me started on this anti-Evolution kick, and how I began to realize how really stupid Evolution is. And, by the way, when I realized it, I had no trouble admitting it, but sure found myself in a lot of hot water since that time for doing so.

When I began my college studies in 1973, the same year as the trip out west when I climbed the mountain, I was already in the Chuck Darwin Evolution Club. Based on counsel from others, I started out with a Business Administration major (that was *not* me!), went through Metallurgical Engineering to Civil Engineering and finally ended up with Biology as my major area of study. I liked biology, and science in general. Astronomy was also my favorite hobby, and is to this day (or better, night).

Oddly enough, it was my college biology texts that really got me to thinking about how silly evolution was, because I began questioning what was being said in those texts, and was not coming up with satisfactory answers. One of my Biology profs and I would get into discussions, and I convinced him to let me give a talk to one of his classes questioning certain things about evolution that we all had learned. Yes, I had come to know God the Creator, but I'm avoiding bringing that into the equation in this book, because my focus is on the fact that, no matter what you believe about religion or spiritual things, or the Bible, the Qur'an, the Bhagavad Gita, or any other book, Evolution is still just plain stupid, and is itself a form of religious indoctrination (with Darwin's book as its bible), and it's time people -- especially scientists – admit it (some already have, but most have not), whether or not they want to believe anything else. **I'm asking the reader to just judge Darwinism and Evolutionary cosmology on their OWN MERITS, and not in relation to any other belief system.** That is why, in the feedback I received on the EvolutionIsStupid.com website, when people so often try to run down the "Bible/Creationist" rabbit trail, I turn it back on them and refuse to go there.

I remember, though, a typical "I'll show him how dumb what HE believes is" response from my professor, which would be one of many such silly challenges I'd receive over the years. He asked, "If God can do anything, can He make something bigger than Himself?" Rather than tripping me up, as he had hoped, I wound up putting him in his own corner with my response: "God, by definition, is infinite, so no, He can't make something bigger than Himself. And that in no way limits His power." The prof, while unconvinced about God, at least respected my answer and had no response.

But it was in phytology, the study of plants, that questions really started coming up. And now we return to explosions again. At a time that Chuck Darwin Clubbers call the "Cambrian" period, there was this EXPLOSION of -- guess what? -- complex life! An *explosion* of the stuff! Virtually nothing before that, in the fossil record anyhow, and the living things that "appeared" in this "explosion" just happened to be very complex. They could in no way be considered "partial" living things that were stepping stones to "more evolved" living things. They were alive, complex, and fully functional, with no further evolution needed.

Complexity. What is it? To a child, tying a shoe might be complex. To an adult that's a simple task. But when it comes to living things, there AIN'T NO SUCH THING AS SIMPLE (that is the technical explanation for it). I like to issue this challenge to evolutionists: Catch an ordinary housefly. Now smash it. There you have all the ingredients for a "simple" housefly.

So, MAKE ONE.

And if that's too difficult, then make a "simple" bacterium. A bacterium is "simpler" than a housefly. So, make one. In fact, make billions of them. Make them reproduce. Have them kiss each other and create men and women and the human sexual relationship while you're at it, just like that silly "science" movie I mentioned at the beginning of this book showed. A bacterium is an incredibly complex creature, and even if we did make one, that would only mean that we were able to DUPLICATE WHAT WAS ALREADY DONE. You see, Clubbers insist that one day man will CREATE LIFE!! WOW! ISN'T MAN INCREDIBLE?! No, he's not. Sorry to disappoint you by THINKING about it, but if man created (note the word) life, he would just be imitating what has already been done, and proving the fact that it didn't happen all by itself with no intelligent intervention. And he's far far far from doing any such thing still, let alone following it up by creating a new kind of brain, which would REALLY be impressive.

CHAPTER XII

Time Is On Our Side, Yes It Is!

The religion of Evolution has a miracle worker, too. It's called "Time," and you just can't get enough of it if you believe that by adding more and more Time, anything can happen. In fact, one of the greatest unspoken beliefs of Chuckie Deeist adherents is...

Chuckie Dee Club Secret #9: IF IT LOOKS LIKE IT WON'T WORK, JUST ADD TIME.

Clubbers KNOW that this stuff about how everything came about and how life evolved is stupid and unprovable. They know it. I know they know it. They just won't admit it. So, in order to get around the fact that they know what they're saying happened could not have happened no way no how, they add Time to it, and then say, "Given enough TIME, anything can happen." Yessir! Like, if I leave a couple of bucks sitting around, maybe in a million years they'll turn into a million bucks (with my luck, though, by then, of course, we will have evolved some new currency system). Maybe my old Rolling Stones albums (which I no longer have in my possession, as my music tastes have evolved, but note the chapter heading here) would have turned into Beatles albums with time! If you just believe in Evolution, you can make Time do whatever you need it to do! Time is the miracle worker of the religion of Evolution.

If you challenge Chuck Darwin Clubbers with the fact that Time itself doesn't really MAKE anything, and that in fact things FALL APART with time, of course they will say YOU are stupid, because all you have to do is look around, and you'll see what Time (the goddess with a capital T) has made, which is everything, or at least that's their belief. They'll tell you that if you put a bunch of monkeys in front of typewriters, maybe in a year they won't type anything, but in a couple MILLION years, they'll hammer out some Shakespeare for ya, yes they will! (Or the lyrics to the average rock song, which is more believable – you know, like"Yeah yeah yeah," or "Hey hey we're the Monkees," or "the pompitous of love" or something like that.)

Of course, that doesn't explain where monkeys came from to begin with, nor how they acquired the ability to bang on a typewriter, which requires a brain, wired up to muscles and bones and fingers and toes, and the ability of that brain to know it's wired up, and know what to do about it. How did that all evolve? TIME did it, of course! Oh, and "mutations" and "natural selection" too. Clubbers will tell you all about mutations and natural selection and all that fancy Clubspeak, but don't DARE challenge them on it. Mutations are the DRIVING FORCE of evolution, man! Uh oh. It's time for THINKING again, and we're not supposed to do that. But we're going to anyhow.

For mutations and natural selection to work, they have to have "something" to mutate and select. It's that question of where the "something" came from in the first place that makes CDEC members squirm, because they have no logical, rational, scientific answer for it, though they spend lots of time, effort, and money trying to figure it out. Notice, though, I didn't say they have no answer. They'll ALWAYS have an answer because all they have to do is remodel the clay and fabricate a new story. It doesn't matter how dumb the answer is, and the dumber it is, the fancier the terminology and scientific jingo for it will be. Remember "abiogenesis"? Well, how about "Punctuated Equilibrium"? Sounds like a Clubber who's had one too many at the Darwin Club Pub, but here's what it's really about.

When Chuck Darwin came up with his idea about how species diversity came about, moving up the ladder from simple to more complex, always changing and adapting, we didn't have very many fossils to determine whether he was right or not, nor was there a known mechanism that could move a creature from "simple" to more complex. If you're in the Club, you believe the fossil record is a "snapshot" of evolution, which supposedly shows that life did move from "simple" upward. But it's a pretty bad snapshot, or at least the interpretation of it is pretty bad, because it doesn't really show anything about evolution; Darwin Clubbers have just INVENTED their nice little stories and drawings of trees and branches and all that, and everyone accepts them, without questioning why the ENDS of the branches have fully formed creatures at them, while the branches themselves show NOTHING in between the "simple" and more complex creatures. There should be MILLIONS and MILLIONS of fossils that show all the stages between molecules and man, and they just are not there, even so many years after Chuck Darwin's time. The fossil record shows a "snapshot" of mostly extinct, fully-formed species. Everything else is in the interpretation, and Clubbers know it.

Because of the *lack* of "transitional" species, clever Club members, always ready to come up with some new fancy cover-up, and led by Evo priest Stephen J. Gould, invented the term "Punctuated Equilibrium" which some called "Punk Eek," to try to make it sound cool. Punk Eek is really nothing more than an admission that the fossil record does not demonstrate Chuck Darwin's evolution. Chuck and his ilk said that evolution took place slowly and gradually over eons of time. But Gould and his pals knew that was not true, and the fossil record showed no such thing, so they decided that evolution took place QUICKLY, in SPURTS, punctuated by long periods of what they called "stasis," when things didn't change much. Well, they figured that took care of THAT! And once again, all the Club members fell in line and paid homage to the Punk Eek god, and once again Evolution was saved from the trash heap.

And although there should be, RIGHT NOW, in OUR TIME, *millions* of plants and animals going through further transitional stages in their ongoing evolution, the fact is, there are NOT. What's there is ALREADY there, and anything that's changing or adapting is not creating anything NEW, it's just acting on what's already there. More on that later, as it's a cogent point that for some reason Chuck Club believers can't seem to grasp.

CHAPTER XIII

Sex Again, and Something Fishy

I think it's time we get back to sex, because Punk Eek gets boring, whether it sounds cool or not. However, the bacteria film (you know, the one where the bacteria were kissing) is a good example of Punk Eek, since there are no transitional fossils shown between bacteria and Bob and Barb, unless you want to believe and imagine that things jumped from bacteria to barracuda to baboon to Bob and Barb. You're perfectly welcome to believe that, even though it's stupid, and you know it's stupid.

But sex is another thing that causes problems for Chuck Darwin Evolution Club members. I don't mean for them personally, though that may be the case, but they can't really explain sex.

I mean, here you have a man and a woman, and they started out as bacteria? Uh, if you say so... (And don't kid yourself. I have *heard* Evolutionist priests like Richard Dawkins say that.) Now, the man is made just right for the woman, the woman is made just right for the man, or at least it appears that way to intelligent people (I don't think I need to get too descriptive here). When a man looks at a woman and finds her attractive, what's going on there? How did THAT evolve? I guess some bacteria were hot, and some were not? Is that STUPID or what?

Then tell me: How did the woman know she had to evolve the right organs to carry and bear a baby, so as to propagate the human race? And how inconvenient is it that humans, more than any other so-called "animals" have to spend so much time feeding and training their kids? (Well, feeding at least; nowadays kids train their parents.) I thought we were supposed to be more ADVANCED? You call baby bottles and diapers ADVANCED? I mean, we were doing doo doo just fine in the forest! Oh, and about those bottles... I suppose EVOLUTION KNEW that women had to evolve breasts for feeding their young, and the breasts knew how to evolve just the right milk, right? And evolution knew the baby would have a mouth that could suckle on the breast to get the milk, and then evolution knew how to make the baby's body assimilate the milk and turn it into bone and other tissue? OH, SURE! That's EASY to believe, as long as you don't THINK about it and believe it by FAITH.

Here's a neat little evolution story that I invented myself (if they can do it, so can I). Think of it like taking building blocks (letters) and making something with them! After you read it, you might think my story came from nowhere and made itself up all by itself, but it really didn't. So here it is.

I can just see it now... Grog and Thogette are your average cave couple a few hundred thousand years ago or so, plus or minus a hundred thousand or so. Why there had to be cave couples, I don't know, but the Club almost always shows them that way in its made-up, imaginary pictures, along with a few little cavebrats running around. Yes, those pictures of cave people are someone's imagination, with no provable basis in fact. Give an artist a bone, and who knows what "pre-humans" will look like when he gets through with it? It's all imagination. Nobody can take one of my jaw bones and tell me what I looked like without having actually seen me, but these guys are dead sure (pun) about what somebody looked like, and where they lived, and how they lived, based on a few fossil teeth or bones.

So, Grog and Thogette decide to have a baby. Where did they figure out how to do that? Did some fish that turned into an amphibian that turned into a chimpanzee that turned into a caveperson show them how? And if fish have babies by laying eggs, then why did that have to change? The fish were doing just fine. In fact, why did they even have to come out of the water? Are they better off now that they're people, on the land, shooting at each other? By the way, fish eggs and other eggs are real "simple" things, aren't they? I mean, what does it take to make an egg that contains all the information to create a new organism? A kid could do it in his spare time when he's done with his building blocks. Just make sure it's a couple million years of spare time, because Time can perform all miracles. The Time goddess made eggs! Thanks, Time! Thanks for bacon, too!

Fish. I guess they were getting tired of swimming around, and decided that it was time to move up in the world. (Some of them did, anyhow. Others were contented to remain fish, conveniently for us, so that we could then invent the story that they are our ancestors). Fish must have known somehow that they'd eventually turn into people, like the Chuck Club tells us happened. Like, umm, we see fish trying to move onto land all the time, don't we? (oh yes, I know – the lungfish. Only problem is, they're still lungfish and they're not wearing business suits yet.) Aren't there still some restless fish out there, or is it that they've seen what we've become, and decided it's best not to evolve into people any more? The fish millions of years ago were different from today's fish. They were more ambitious, not like today's apathetic, X-generation fish who are just swimming through life looking for meaning in an ocean of salt and seaweed.

(An aside here. Do you know what's really gross? We *eat* fish! We EAT OUR ANCESTORS! YECH!! Does the Chuckie Dee Club ever think about this stuff? I have actually had Evodelusionists mock me for saying that. Listen: If A equals B, and B equals C, then A equals C. Therefore, if you believe your ancestors were fish, and you eat fish, then you're eating your ancestors (others of whom inhabit your colon, if you hadn't thought about that one). Don't try to get around it. Some of our ancestors we eat, and the rest we swat, poison, shoot, or put in zoos, or whatever. Or we insult them, like when we say, "You're acting like an ass." Donkeys are not standing around wondering where they came from, and inventing stories that they were once fish. They're too smart for that. You're acting like a *human*. Trust me.)

So fish moved onto land, eh? If you watch Evodelusionist "science" propaganda movies, you'll see that they did. And science films do NOT lie. Nothing to it! All the fish had to do was grow limbs, change their respiratory systems, change their diet and digestive systems and find new food sources, change their skin and eyes so they could see and live outside the water and not dry out, change the direction of their motion from side-to-side to forward and backward, try not to get sunburned or freeze once they left the water, change from cold blooded to warm blooded. NOTHING TO IT!!!! All it took was TIME, and there you go!

Have you ever seen what happens to fish out of water? They DIE. However, it should be OBVIOUS to you as an Evolutionist that, with MILLIONS OF YEARS OF TIME, all the changes that fish needed to live out of water took place despite the fact they were dying, and soon enough, they turned into people. STUPID, is what it is. Darwin Clubbers KNOW it's stupid. They just won't admit it.

Perhaps the Mother of All Marine Animal Stupidity (MAMAS) is the WHALE evolution story. To be a sworn member of the Chuck Darwin Evolution Club, you have to believe that whales started out in the water, left the water, and then *returned* to the water! What DUMMIES! They could be PEOPLE by now, fighting for whale rights for their ancestors who didn't evolve. But NOOOO! They had to go BACK to the water.

And those blowholes are another issue. Members of the Chuckie Dee Club don't care how whales breathed before they had those blowholes. They prefer not to ask those questions. We just KNOW the blowholes evolved with millions of years of Time and we'll eventually find all the transitional forms, and that's all we need to know. There aren't any "transitional" blowholes, though there are a number of traditional blowhards in the Club!

CHAPTER XIV

Dinosaurs Take a Bite Out of Darwin

Speaking of whales, they're pretty big critters. And guess what? Chuck Darwin Club members know WHY they're big: It's because they had to OUTGROW THE COMPETITION! Bigger is better, according to Evodelusionists, and somehow Evolution knew that, so it made things big so they would survive better. That's why dinosaurs were big (even though they were not born big, but let's not bring up too many problems here, ok?).

Doesn't it make sense that bigger is better for survival? (If the average overweight American is any indication, we're doing just fine in that area.) No, it doesn't make any sense whatsoever. In fact, it's STUPID. The reason it is stupid is because there are millions of living things that are NOT big, and have NOT gotten bigger with time, and they've survived just fine -- even better than some big things. Most of us are more afraid of a virus or *E. coli* than a whale or an elephant! Viruses are pretty tiny. Real tiny. But if you're in the Chuck Darwin Evolution Club, you have to invent stories about why things are big, or why they have such SHARP TEETH (my, T-rex, what *BIG* teeth you have!). If they have sharp teeth, then they must have been FIERCE WARRIORS, that's for sure! Even though there are animals with big, sharp teeth nowadays that are NOT fierce warriors. But images of complacent vegan dinosaurs don't sell books and movies. Dinos tearing each other apart do.

So, as the dinosaurs' teeth got bigger, they needed bigger bodies to fend off predators (and also for cosmetic reasons), right?   Or is it that as their bodies got bigger, Evolution knew they needed bigger teeth? Like Evolution was just sittin' there one day, and Evolution says to herself, "Evolution, you know, you have to make dinosaurs bigger, and their little teeth too!" (that's a play on a line from the Wizard of Oz movie, in case you didn't know). And it was so. I do think a little dino with gigantic teeth would have been cool, though.

Why Evolution didn't make bacteria bigger, we'll just never know. I can see a big, hotdog-shaped bacterium with long fat legs devouring a city in its struggle for existence, can't you? So why did Evolution make dinosaurs big, and bacteria small? Of course, bacteria are all over the place, and dinosaurs are gone. So Evolution was not very smart there, I guess. But despite this, Chuckie Dee Clubbers will insist that the dinosaurs got bigger to be able to survive better. Oh, now, CDEC members will attack me for all this, and they'll say, "Evolution does not think or have any direction."

Well, then DON'T PRETEND IT DOES.

Don't come up with excuses for why some things got big, and others stayed small. We don't need your explanations if there's no direction or thought involved. It was just random accidents, some of which got lucky, so don't bother trying to figure it out or pretend like you know what you're talking about. You don't.

Which brings to mind "the mind." Let's suppose I say 2+2=5 and you say 2+2=4. Are you right, or am I right? You'll say that you're right, right? But if Evolution is true, suppose it's just that the chemicals that evolved into MY brain, did not evolve the same as the chemicals that evolved into YOUR brain, and we just understand the world differently. Maybe

I'm counting the numbers and symbols in the equation (3 numbers and 2 symbols) and you're adding two and two together. Or maybe I just don't want to see things your way. Who are YOU to say I'm wrong about ANYTHING? Once again, Evolution leads to STUPIDITY, because if your brain is just a bunch of evolved chemicals, then nothing has any meaning in the end. It's just the way your chemicals happen to perceive things, and mine perceive things differently. SO MIND YOUR OWN BUSINESS and leave me alone!

Or maybe I came from a "parallel universe" where 2 + 2 really does equal 5 and you just haven't figured out how. Yes, parallel universes, baby universes, multiverses – all Evodelusionist science fiction inventions with no basis in observed reality, yet accepted as realistic possibilities by our modern intelligentsia, but it's not ok for me to say 2 + 2 = 5?

We're getting out of control here. Let's get back to dinosaurs, because they're fun, and kids like them, especially because they're extinct and cannot eat them for breakfast like kids eat certain dinosaur cereals (there's teaching your kids to eat their ancestors again; I've seen dinosaur cookies and other dinosaur foods too. We just can't let a sleeping dino lie, can we?).

What happened to the dinosaurs, anyhow? Well, if you belong to the Chuck Darwin Evolution Club, you have OVER 100 POSSIBLE ANSWERS to that question, and more are being added all the time. The best so far are that they drowned in their own dung, or plants died off and their diet changed dramatically, so they died of constipation. Talk about two extremes! What a bunch of coprolite! Perhaps nowhere is the stupidity of Evodelusionary scenarios more evident than in answering the question of the demise of the dinosaurs.

One of my favorite dumb Evolution things is when Darwin Clubbers get philosophical and tell us that somehow knowing how a dinosaur lived "65 million years ago" or how a star formed elements and then exploded tells us something about ourselves, our place in the universe, and our future. The hogwash meter about breaks over that one. WHAT FUTURE? If all you are is evolved scum, then that's what you're going back to. You'll be feeding whatever's evolving after you've come and gone. What wonderful HOPE! What a sense of PURPOSE! Now I KNOW what my future holds, and..., and..., well, SO WHAAAATTT??? If I came from scum, and I'm returning to scum, then I have no real purpose, no real past, and no real future, so don't try the philosophy stuff on me, because it's STUPID to tell me about my "place" in the universe in the grand scheme of things. Instead, please tell me:

How did nothing become something and turn itself into everything?

Maybe you've heard about the Asteroid? OF COURSE, YOU DUMMY!! The dinosaurs were killed off by an EXPLOSION!! You were expecting something less? Have you forgotten about the coolness of explosions, and how they made everything? Well, now they're destroying everything! An Asteroid smashed into the Earth, and somehow it chose to mess with the dinosaurs, while leaving lots of other things alone. That's what the dinos get for increasing in size. See, things never work out the way you want them to. Here were the dinosaurs, dominating the Earth (supposedly) and puffing up their chests with pride at how big they were, and along comes this Asteroid and shows them a thing or two.

Now, the REALLY weird thing is that, most of the remnants of dinosaurs, mainly bones, are BURIED IN SEDIMENT. Like they died in, you know, a FLOOD. Like that one on Mars, where there's no evidence of water that could cause a flood. But no, it had to be an Asteroid, because if we say it was a FLOOD, that's too much like admitting that some other belief might be right.

CHAPTER XV

Fossil Foolishness

Let's go back to my mountaintop trip for a moment. Along the way, we also stopped at Dinosaur National Monument. There you can see an entire side of a mountain that has been cut away to reveal dinosaur bones that were buried in *sediment*. In fact, if they were not *quickly* buried in sediment, we would not have them to gawk at now. The Chuckie Dee Club will tell you that sites like this were the result of a "local" flood, or dinosaurs and other creatures being swept down a raging river. I have lots of articles or have seen in evolutionary books where they say this fossil deposit or that was the result of a "local" flood or a raging river. Some of the deposits have hundreds, thousands or even millions of animal fossils in them. That's an awful lot of stupid animals (don't forget: stupidity had to evolve too, so I guess maybe these really ARE transitional forms!). Unlike people, however, animals must have gotten smarter over Time because, well, I've seen LOTS of local floods in my lifetime, and not one of them produced any fossils, let alone millions of them.

But Evodelusionists don't care about that. They have to have SOME kind of explanation, so why not invent a dumb one like that a bunch of animals were buried in a "local" flood and turned into fossils, even though that would never actually occur in the real world?

Chuckie Dee Club Secret #10: WE KNOW FOSSILS COULD NOT HAVE FORMED GRADUALLY, AND WE HATE TO HAVE TO ADMIT IT.

Fossils are the remains of creatures -- mainly marine ones -- that were buried QUICKLY, and beneath LOTS of sediment, which is now wearing away, exposing them. Most fossils are only the remains of hard parts, like bone and shell. I have seen a picture in an evolution book that *really* exposes (pun intended) just how stupid Evolution can be. It shows a "prehistoric" caveman who is circumcised. You should not have to think more than a second or so to realize how ridiculous that is. It's pretty unlikely that a fossil was found with corroborating evidence for that, and circumcision is a religious practice that's only a few thousand years old. But it goes to show how much artistic IMAGINATION and BIAS is involved with this stuff.

Have you noticed the "**Oldest Man Game**"? If not, then you've not been reading enough Evolutionary propaganda. You see, there's this contest among paleontologists to see who can find the oldest human. So one paleontologist finds a few little bones (usually in some remote area where few others can go – kinda like where UFOs are always spotted) and says, "My human ancestor is *two million* years old!" Then the next paleontologist comes along (and you can be sure the news headlines will say, "New Find Rewrites Human History!") and he finds a bone or two, uses an artist's imagination to decide what "it" looked like, and "Eureka! *Mine* is **THREE million** years old!" Well, the other guy's not about to be outdone, so he finds another bone or two, has an artist draw a half man, half ape creature that looks older than the other guy's, and "Aha! *Mine* is **FOUR million** years old!" Pretty soon mankind is going to be older than the universe, because they're not about to give in to the other guy, that's for sure! It's the paleontological pre-human dating game! And you are just the dummy sitting in the audience accepting what the "experts" are telling you.

Getting back to dinosaurs once again, and being buried in sediment... Where did that sediment come from? Well, the oceans of course. A few years ago I attended a talk given by a local geologist about the oceans that covered Wyoming in the past, but, being a member of the CDEC, the speaker didn't tell us where those oceans came from. You see, the ocean didn't cover Wyoming just once, but TWICE.

Well, if that's the case, there should be lots of fossils under the present ocean, whatever it's covering, shouldn't there? But I don't know of one that's been found about which it can be said that it formed *IN SITU* beneath the ocean, despite all the digging on the sea bottom. Have any dinosaur fossils been found under the ocean? How about any enormous fossil clam beds? Shouldn't there be literally countless billions of fossils beneath the ocean beds, or at least the continental shelves? How about modern lakes? Supposedly birds, fish, etc., were fossilized in ancient lakes when their remains sunk down and were covered with sediment. Why don't we see that happening NOW? I mean, right now, we should be able to dig in any lake sediment and find these fossilizing creatures. We don't. For the simple reason that before they could become fossils they'd be scavenged and decay. Clubbers know this full well. They just won't admit it.

Oh, I know, the Club has an excuse or a fabricated story for all this, but they have not heard of my "**Elevator Theory**." You see, according to Evodelusionists (maybe that should be Evo-deluge-onists for the time being), the land was once under the ocean (hence, all the marine fossils we find, even though none are forming under the ocean now), and then the land rose up, and began draining off. The Grand Canyon is supposed to be an example of this. But once again, the STUPIDITY METER is going to break. Why? Because there's no way that the strata of the Grand Canyon could have been laid down over "millions of years" and then lifted up for "millions of years" and then eroded for "millions of years." That's what I call Elevator Geology and it's one good reason that Evolutionist scientists are constantly arguing about the age of the Grand Canyon. Up, down. Up, down. And without any rhyme or reason other than inventing another just-so story to try to explain the Grand Canyon and other similar formations worldwide, with no real basis in fact.

The strata of the Grand Canyon were obviously laid down contemporaneously (in the same time period), not over millions of years. That's why the science gods can't agree on an age for the Canyon (like they can't agree on how the dinos disappeared). Because they can't see the data for the strata. There are pretty well-defined breaks between the strata, showing they were laid down around the same time period by large hydrodynamic forces, differentially depositing rocks, stones, granules, sand and silt, not gradually depositing them in quietly flowing streams, or lakes, or local floods. In an enormous, worldwide flood, (as opposed to a local one) the material would be all mixed up and turbid, and then when it started to settle out of rapidly-flowing drainage activity, it would form strata as the different size and weight material was laid down (this HAS been demonstrated scientifically – it is FACT).

If the material had been laid down gradually by relatively moderate amounts of water, it would not show the sort of definition (strata breaks) in the quantities (extremely thick homogeneous strata) in which we find it all over the world. Any kid can fill a jar full of water with a mixture of dirt, sand, and stones, and shake it up, and see that it forms strata and doesn't take a million years to do so – it's all about the right mixture and the movement of the water, and has little to do with Time. I can go to a local area right near where I live and see strata that form each winter as snowmelt causes the water to ebb and flow. That shows that strata do not take a long time to form, but for the strata to be homogeneous and laid down in the huge thicknesses and over enormous areas like they are all over the world would require a mixture of different-sized objects, from large stones on down to grains of sand, and an enormous amount of moving water.

So a kid can see this, but if you're an Evolutionary scientist with ten PhDs and belong to the Club, for some reason you can't see what a kid can see. Well, actually, I believe you CAN see it very well, but you just won't admit it. After all, your other Club friends might stick their tongues out at you and call you a religious fanatic or something because you believe in a worldwide flood. Of course, THEY are not religious fanatics, even though nobody ever saw happen what they say happened, namely that there was a Flood on MARS! Don't forget that! *Their* religion tells them so!

CHAPTER XVI

A Body of Damaging Evidence

Enough of dinosaurs and fossils. Let's talk about the human body. Now, we all know that the human body started out as a one-celled animal (say, a bacterium or some other slimy critter), that grew into an amphibian, which became a chimpanzee, which became Joe and Jane Person, right? Do I sense the STUPIDIY METER starting to move? You bet. Because that is ridiculous, and the Chuckie Dee Club knows it. They just won't admit it.

We'll start with digestion as one example of how stupid it is to believe Evolutionary mythology.

First you have to have a food source that just "happened" to evolve the right foods for your nutrition, and just "happens" to be edible. Evolution took care of that. Then you have these hands, you know, that can be used to harvest and prepare the food for eating. Evolution evolved them. The hands then bring the food to the mouth by way of arms which just happened to evolve in a way that they could help the hand reach the mouth, which just happened to have evolved the right size opening (and ability to open) to receive the food, and has a tongue inside with taste buds that evolved and connected themselves to the brain so they could help you decide whether the food is good or not, teeth of the correct size, form and location to grind the food down, saliva to lubricate the food for passage down the esophagus to begin the digestion process, muscles and bone to operate the process of chewing and swallowing, an esophagus to transport the food to the stomach, a stomach with the right structure and juices to further break down the food, which Evolution connected to intestines that begin absorbing the right nutrients from the food, and a bloodstream to transport those nutrients to every cell in your body, with a liver and kidneys to filter out toxins and regulate water, muscles to push the digested food through the body (and you don't hardly know it, except for an occasional cramp and maybe a bit of flatulence now and then), and finally an orifice where the digested food can be eliminated and become a part of the nature from which it came, starting the process all over (sorta like "Slim.")

And ALL THAT HAPPENED ALL BY ITSELF WITH NO PLAN OR DIRECTION? Evolution is just AMAZING, isn't it? Time sure worked a few miracles there, didn't she?

Well, the notion that all that came about on its own is STUPID. And we've only touched on the digestive system. We haven't even mentioned the nervous system, or the BRAIN that MAKES IT ALL WORK. HOW DID THAT HAPPEN? Did Evolution wire everything up just right, and if so, how did it do that? All you have to do is stop and think about what the Chuckie Dee Club preaches, and if you don't arrive at the conclusion that stupid arrogant stubbornness is involved in the continued propagation of this inanity, then you should have your mind checked, if it hasn't been checked at the door already.

***

Here's something else to think about (off-topic for a moment): People still live in caves around the world, right? But that doesn't matter to the CDEC because nowadays upscale cave dwellers have computers, TVs and air fresheners in their caves, which they did not before, so obviously they're more "advanced" than those "prehistoric" dumbos were. What am I getting at? This: Why are THEY not called "cavemen?" It's because that would mess with the evolutionary timeline, that's why.

And how about "Stone Age people?" There are still people around who use stones as tools (I've done so myself on occasion), and even more people did so not very long ago in history. Why is it that they are always COMPARED TO "Stone Age people?" They ARE Stone Age people. NOW. TODAY. Not hundreds of thousands of years ago. NOW. So why not get off the "Stone Age" kick and just admit that stones have always been a part of the history of mankind and have nothing to do with a "Stone Age"?

Now I need to get onto another off-topic pet peeve (no pun intended), namely, *animals* supposedly using "tools." That's a bunch of anthropomorphic baloney if ever there were any. A monkey puts a stick in a hole in a tree and pulls out some ants and suddenly one of our "ancestor-like" creatures is USING TOOLS!! WOWEEE!!! The next thing you know, monkeys will be building automobiles, that's for sure! Or an otter uses a rock to break open a shell to eat the contents. It's using A TOOL!!! PROOF OF EVOLUTION, YESSIR!!

Well, sorry to disappoint you Evvies, but the animal is *not* using a "tool." It's using a stick, or a rock, LIKE a tool. But it's *not* a "tool." You're just adding that human dimension to bolster your Evolutionary mythology, and nothing more. The leap from using a stick to get at ants in a tree hole to using a drill to put holes in a two-by-four to build a house is so astronomically distant as to make any notion that animals use tools (or speech or whatever else your imagination lets you fly away with) utterly laughable. But we're not laughing at it. Intelligent people are preaching it. Intelligent people believe it. **We should be LAUGHING at this stupidity.**

***

Back on the topic of what makes us human, I'm afraid we hit a real problem for Clubbers when it comes to communication. In order to have communication, you have to have a codified language (that has some assigned meaning behind it), a transmitter, a receiver, and a way to decipher the code and put it into practice. With speech, the transmitter is the voice, the code an alphabet, organized into words, the receiver is the ear, and the putting into practice is done by motion or other means. To say all that evolved is... you guessed it. Say the word yourself. To believe this stuff, you have to accept the following stupid scenario:

The body knew it needed a mouth. The mouth knew it needed vocal chords. The vocal chords knew they needed air. The brain knew it could come up with a coded system using air, vocal chords, and a mouth. The ear knew it could hear the noise propagating through the air from the voice box. The ear knew just how to arrange itself (what parts it needed) to transmit all of that to another brain. The other brain knew just how to decode the message it received. The other brain then knew just what sort of body it needed to put that message into effect.

To say that that whole system of communication evolved on its own is ridiculous. There's not a shred of evidence to support it; it's an illogical, irrational, fanatical evolutionary religious story. And we haven't even gotten to the EYE yet!

How did an eye "know" it could see light? How did Evolution figure out how to get the eye to see light? *How did Evolution even know light existed and COULD be seen?* It DIDN'T. And there should be fossil waste landfills everywhere that are full of evolutionary experiments that failed while Evolution was trying to figure out how to make a mind-numbingly complex human eye that could utilize light (not to mention all the other countless Evo "mistakes" that should be in those landfills). Clubbers know without a shadow (pun intended) of a doubt that the eye did not evolve despite their iconic fictional "transitional" images showing an imaginary transformation from an light-sensitive spot to a human eye. But the Club's rules say you have to believe the eye evolved all by itself and hooked itself up just right to your brain if you want to be a member. No matter how stupid the belief in Evolution, they're just not going to admit it.

CHAPTER XVII

Millions and Billions and More Billions

Back to the mountaintop trip we go. While in Rocky Mountain National Park, we joined in a tour hosted by a park ranger. I'll never forget how he picked up a seashell off the top of a mountain, held it up and said, "Millions of years ago this mountain was covered by a sea."

Ok, so a seashell lasted millions of years on a mountaintop? How come I've been going to the seashore all my life, and each year there are new shells all over the place, and the old ones are gone? But this one -- and BILLIONS of others, according to Chuck Clubbers -- lasted MILLIONS OF YEARS!!!? Whoa baby!!! And the mountains were beneath the sea, got lifted up, rained on, snowed on, blowed on, beat on by the sun, sand blasted, and here's a shell that withstood all that for MILLIONS OF YEARS. Yeah, right. Oh, that's right, it was BURIED all that time. Sorry, I completely forgot! So we'd better do a REAL science experiment then. Go get a clam at your local seafood store. Now bury it. Say, a foot deep if you wish. Be sure to mark the spot. Now, wait five or ten years. Then go look for it. If it turned into a fossil, give me a call. If you can't find it, don't bother.

But at the time, I believed all that evolutionary baloney. That is, until I started to actually THINK about it, instead of just believing every story I was told. Then I found out another fact: Almost all fossil clams are found CLOSED. Ever been to the seashore and seen dead clams? You know what happens when they die? That's right, they OPEN UP. So, all over the world, we have fossil clams which show they were rapidly buried, not buried nice and slowly by sediment being carried by waves or a stream or a local flood. And not one or two buried rapidly, but BILLIONS, and from what I've read they're even found on top of Mount Everest! I don't care what you believe about how mountains formed; NO clam fossil – buried or not -- could survive *millions of years* of environmental abuse and be sitting there waiting for you to pick it up. I have rocks, brick, mortar, etc., all around my house as do others, and in a FEW YEARS I can see evidence of weathering and wear on them, and these guys are telling me these clams and other fossils have hung on for MILLIONS of years? I DON'T THINK SO. Some of the mother-of-pearl is still even on them.

How about amber? All sorts of critters have been found in amber, from frogs to ants to bees to plants, and so on. And guess what? They look JUST LIKE frogs, ants, bees and plants, and so on. They're not somewhere on the way to *becoming* those things; they ARE those things. Now, ain't that a coincidence? They haven't changed a bit in all those "millions of years?" Well, that's because they haven't been *around* for millions of years. And Chuckie Dee Clubbers know it, but...

Chuckie Dee Club Secret #11: NEVER ADMIT THAT IT "LOOKS YOUNG," EVEN IF ALL YOUR SENSES TELL YOU IT IS.

In other words, even though they can see it's a frog, or an ant, or a bee, or a plant, and that it hasn't changed a BIT in "MILLIONS OF YEARS," there is NO WAY they're going to admit that it just MIGHT be LESS than millions of years old, because that goes against the party line of their religious myth. Just imagine: In the same time that fish were turning

into people, a lot of these things DIDN'T CHANGE AT ALL!
Isn't that AMAZING?! Of course it is, if you're a member of
the Club.

One smoke and mirrors tactic they use to trick you into
thinking it's something other than what it looks like is to assign
it a fancy italicized Latin name. So, instead of just saying they
found a FROG in the amber, nooooo. They found "***Protofrog
slimeballensis***" which is supposedly evolving into a modern
frog. And YOU are impressed, and suckered into believing
it's something that just "looks" like a frog but, by golly, it's an
evolving frog!

***

MORE TRAVELS WITH JOHNNIE

Let's take a brief trip to Albania, where I went in 1993, and
was asked to speak to a group of mainly young people
(college age or thereabouts). Of course, I spoke on evolution --
that is, what's wrong with evolution. I brought up the famous
"peppered moth" which is one of a handful of so-called proofs
that the Chuck Club holds out to the public, and the public
accepts it either because they don't know any better, or they
just don't care. I pointed out how the peppered moth story
had nothing to do with the Evolution myth, and a college
student came up to me quickly afterward. "We just studied the
peppered moth last week!" she gushed, "...and I'm glad to
know why it's wrong, because I didn't believe it anyhow!"

Ahh, if only Chuckie Clubbers were as honest and fearless as
that young girl. Peppered moths never "evolved." The whole
story was contrived and had nothing to do with moths
changing INTO anything other than moths. In a nutshell,
darker ones traded places with lighter ones as pollution
supposedly darkened the trees where they hung out, then the

lighter ones took over later, after the Industrial Revolution pollution started subsiding (that's if we can even believe *that* part of the story). Nothing whatsoever to do with where moths came from in the first place, or how they "evolved" color, or anything else having to do with an increase in complexity or a moth changing into a sloth. The moths were still moths. Nothing new "arose" in their makeup.

Move on over to Russia, where I had the opportunity to do brief talks about evolution in eleven orphanages and other places in the province of Udmurtia. Again, the same response. People were so happy to hear that the Evolution myth they had been FORCED to believe with no alternative was just a hocus pocus pack of lies. They KNEW it was a pack of lies, but could not challenge it openly, and just lived with it.

All over the world, people are beginning to awaken and recognize that they can't stay in their shells about this forever. You can try to clam up, but this theory is dying, and when it finally breathes its last, the shells are going to have to open up. It's all about WHO WE ARE, WHERE EVERYTHING CAME FROM, and WHERE WE'RE GOING, so it's not a minor issue, like some people will say, which gives them the opportunity to brush it off without having to THINK about it.

Millions and Billions. What about the rocks? Aren't they millions and billions of years old, and doesn't that PROVE Evolution? Sorry, Chuckies, but it doesn't, and you know it doesn't and you KNOW you cannot PROVE the age of ANY rock. I used to be in a rock group, many years ago when I had more hair. In fact I used to perform some songs by the Rolling Stones, so that kind of shows how old I am but it won't help you much with the age of rocks -- just the age of rockers (or ex-rockers in rocking chairs). The age of rockers can be determined, because we have a written history of their lives. The age of ROCKS can't be determined. Some rocks that have formed recently have given radiometric dates of millions of years, but Chuck Clubbers don't want you to know that. We CAN'T accurately date rocks, no matter what you're told. There is NO rock that anyone can hold up and say, "This rock is one million, two hundred thirty three thousand, five hundred twenty four years, and 67 days old." That's STUPID, right? Ok, go ahead and attack me now. Sure we can't be that accurate! Oh yeah? Why NOT? Don't tell me about a rock that's a million plus or minus 100 thousand years old. I'm not impressed. Does anyone ever stop and think what a huge margin of error these storytellers allow themselves?

Well, it's stupid to believe anyone knows the age of rocks, too. The whole age thing is a sham. The Chuck Club geologists know it, but they have you fooled into thinking they are the coolest rock group going, because they can look into their rock crystal balls and tell you everything that happened millions of years ago, and how things looked, and how people lived, and what animals did, and what the weather was like, ALL FROM READING ROCKS! I mean, don't you wish they could tell us what things are going to be like in another million years? Or maybe just *tomorrow* would be helpful.

Now, I'm not saying you can't know anything from reading rocks. You can know how they're formed, you can know what they're made of, and how they may have changed (metamorphic rocks) and even how they weathered, but *only within the confines of here and now* – that is, your *present* observations. You can NOT know ANYTHING about a rock "millions of years ago." It's all based on GUESSWORK and STORYTELLING, and they KNOW that, but won't admit it.

Turning once again to the professional geologist I recently heard speak, as she showed a slide presentation of many of the interesting geological formations around where I live, inwardly I was cringing, but out of politeness I didn't say anything. Then she left quickly afterward so no one could ask questions. But there were at least two things that stuck out above the rest as prime examples of the kind of thing I hate that Evolutionists get away with.

The first was when she showed a slide of a rock formation and said, "This formation is about 2.6... oh, let's say 3 *billion* years old!" Well, were we supposed to all oooh and aaah, like we'd just seen fireworks, or what? It may not immediately strike you as to what's so funny about her statement, but with a wave of her magic geology wand she added FOUR HUNDRED MILLION YEARS to the age of the formation. And who's gonna notice? What's the big deal if you just decide it's 3 billion, instead of 2.6 billion? In the grand scheme of things, is it going to make my day any more enjoyable or miserable? Of course not, so why should I bother challenging it? Because that WOULD make my day more miserable, because the Clubbers would get out their swords and come after me if I challenge their billions and billions and millions and millions. I am threatening their goddess Time.

Evodelusionists throw around millions and billions like gamblers who have too much money on their hands. If a million doesn't work, maybe a billion will. I did an Internet search one day for the age of the universe, and came up with between about 8 and 20 BILLION years. Now THAT'S exact science for ya!! Yessir! When I was a kid the universe was supposedly about 5 billion years old, if I recall correctly. Then it went up to about 20 billion, and now it's back down to about 12 billion years. Is there some reason they can't make up their minds? I mean, if it's so OBVIOUSLY old like they say it is, why can't they come up with a specific age? Like that the universe is 10,324,596,144 years old? Now THAT would be impressive! Not a span like 12-15 billion years. That's a margin of error of three BILLION years. Doesn't that tell you Evos something?

The second thing our geologist did was tell us about a
formation that was MISSING millions of years of strata.
That's because she'd bought into the infamous "Geologic
Column" fantasy that doesn't exist anywhere in the world in
the same way it appears in books. So there were millions of
years of strata that were supposed to be there, according to the
mythology, but they were not. She had no explanation for
that, but of course we all have faith that it will some day be
explained. There was no question, of course, as to whether
the whole "Geologic Column" scheme might just not be true.

The bottom line is this: Whatever the actual facts show, no
matter how contradictory, we can't question the myth.
Rather, we just adjust and either ignore the facts or fit them
into the myth with some fictitious story that "might have" or
"could have" happened. If anyone else did that, especially a
"religious" person, they'd be mocked and condemned.

CHAPTER XVIII

So, Are You a *REAL* Scientist?

Science has become an icon in our day. I know full well that
members of the CDEC will attack me for this book. They'll
call me stupid, say I don't know what I'm talking about, say
I'm not a "scientist" (whether or not they know my credentials
and experience), or that I'm not a *REAL* scientist; they'll try to
find out what I believe and pick it apart, and just basically do
what they can to suppress the ideas in this book. Why?
Because they have to protect the Club and their Evolutionary
religious myth. They can't let on that maybe they agree that
Evolution is stupid, because that's like saying *they* are stupid
for believing it. Well, that may or may not be true. Most
people believe Evolution not because they themselves are
dumb, but because they trust the "experts" who are feeding
them evolutionary fast food, and so they don't bother
questioning whether or not it's true. I mean, after all, if Joe
Club Member has five PhDs, he MUST know that Evolution is
true, and WHO AM I to question him? Or, as I've said before,
they have some vested interest in the theory.

I have news for you. It's the FACT of those PhDs and other
highfalutin' degrees that renders many high and mighty Club
members unable to admit that Evolution is baloney, because
they belong to a Club whose members will shun them if they
admit it's just that. It is time for you PhDs and other
intellectuals to admit that Evolution is false, it is a lie, it has no
substance, it is misleading, it is NOT science, it is NOT
NECESSARY to ANY branch of science, it is a belief about

who we are, why we're here, and where we are going, and
nothing more. It is a RELIGION. I told you I don't care what
else you believe, and I am *not* bringing my own beliefs into
this. I am just calling for you to STAND UP and show some
GUTS and admit that evolution is a FARCE. And if you don't
like capital letters, that's too bad! I AM yelling!

At the beginning of this book I quoted the famous astronomer
Giovanni Schiaparelli, who referred to those who **"use their
own imagination as an instrument of research."** If that
doesn't describe the average Evolutionary scientist, nothing
does.

Evolution is NOT science. And here are just a few reasons
why:

- **The Big Bang exists only on the blackboard, and in
  the imagination.**
- **The formation of the universe exists only in drawings
  and the imagination.**
- **The formation of the solar system exists only in
  drawings and the imagination.**
- **The evolutionary beginnings of life on earth are
  products of the imagination and nothing more.**
- **The so-called "soup" from which life had its
  beginnings never existed.**
- **The rise of life from non-life is not possible, and is a
  fairy tale.**
- **The increase in complexity in living things is also a
  fairy tale, and product of the imagination; the actual
  evidence shows the reverse.**
- **Mutations, which supposedly drive evolution, do no
  such thing, and you know it.**
- **There is no way that the diversity of life could be a
  product of chance, and you know it.**

- **There is no way that human beings could have come from single-celled organisms.**

And you know it.

That's just for starters. Let's take another look at mutations, which are supposedly the champions of evolution, together with the miracle worker, the goddess Time. If mutations are such a great thing, and so vital to the advancement of life on Earth, THEN LET'S ALL GO TO CHERNOBYL FOR VACATION, where we can enjoy some good exposure to radiation, which will mutate our germ cells and cause our offspring to be better evolved than we are. Right? I see Chuck Club members squirming a bit. Why? Because you know that mutations are almost always HARMFUL. You also know that mutations don't ADD anything to the genome, but use what's ALREADY THERE in the genetic makeup of living things (for some reason that's a really hard concept for Evos to grasp, which is why I have to keep repeating it).

The first mutation... what was it? Was there some blob that you would call a "sort-of-cell" floating in the ocean till it was hit by lightning or a cosmic ray and started the ball rolling toward every life form that exists and ever existed? There has to be something to mutate in the first place before mutations can make everything, right? Your problem is you have no idea what that something was, nor will you ever. But because you belong to the Club, you can't bring yourself to admit it.

Since I was a little kid, perhaps from the time I saw Sputnik coming up like a dim star on the horizon from a park near the Philadelphia airport, or the first time I looked at the moon with awe, I have loved to look up into the sky and wonder about what's out there. I love astronomy, and have been a big fan of the space program from its inception, too. That's why the whole SETI thing rubs me raw. They can look out into space and NOT see design, look at the hands and brains that made radio telescopes and the incredible electronic components behind them and NOT see design, but when they supposedly find a SIGNAL FROM INTELLIGENT BEINGS they think they'll recognize design.

Here's a message from an intelligent being:

"HELLOOOOO!!! TRY LOOKING AROUND YOU, ETers!!! The question isn't, 'Is anybody out there?' but rather, 'Why can't anybody see design in a DNA molecule, or a baby being born? Anybody home???'"

End of message.

But it's not just SETI. It's the "billions" thing again. Here's the famous Chuckie Dee Astronomer equation:

Stars are billions of light years away. That proves that the universe must be billions of years old. Which proves that life evolved on Earth. IS THAT CLEAR? Don't you see the connection? You don't? Good. That means there's hope for you.

But if you don't see the connection, then you've forgotten about goddess Time. She can do ANYTHING. Only non-Clubbers cannot see the connection between how far stars are from us and the existence of life on Earth. If you don't see it, you're not a REAL scientist.

Here, let me explain it this way. In 1976, I rode a bicycle across the USA. About 2500 miles or so in 45 days. So, if there are 2500 miles and 45 days between Philadelphia and San Francisco, that PROVES that grizzly bears evolved in Alaska! You see? You don't? Then perhaps you are on your way to the freedom of chucking Chuck Darwinism and the religion of Evolution. Or perhaps you've chucked Chuck and already experienced the freedom from being shackled to a lie. If so, good for you, but now you need to go back and tell the Club members that you no longer accept their rules, because Evolution is stupid, you know it's stupid, you can admit it's stupid, and it's time to move on and for them to do so too.

By the way, I believe one day they WILL find life in outer space. But it won't be real. It will be a fabricated report, and all the deluded Evolutionists will fall in line, ready to believe it. Then the truth will be revealed. I won't get into that further here. It's just something I believe humanity is being set up to fall for. (See the fascinating book on UFOs that I've included in the chapter on Further Reading.)

In 2005 I started the website, www.EvolutionIsStupid.com, to continue this discussion more in depth. When I decided to publish this book after numerous requests to do so, and since I had edited it quite a bit from the original prior to publication, I removed the main text from the site. Despite many attempts by my detractors to corral me over the years, I did not to get into my personal beliefs, at least not yet. That is not the point I desired to drive home. I just felt that it was time to stop pussyfooting; time for those of us who have rejected Evolution to stop having to defend ourselves; time for Chuck Clubbers to own up to the fact that Darwinism and the whole of evolutionary cosmology is kaput -- bankrupt, sunk, and dead. It is useless to the advancement of science or society or anything else. We didn't need it to get to the moon, don't need it to advance medicine, didn't need it to build the Hubble telescope or any other instrument, don't need it to extract energy resources or teeth, don't need it to make law, don't need it to cook dinner, don't need it to advance automobile or computer technology. We simply don't need Evolution for anything other than to substitute it for some other belief about origins, purpose and destiny that we don't find palatable.

Evolution is a nice, non-threatening religion, and hey, eventually we'll even be able to CONTROL evolution! We'll be our OWN "gods!" We can do whatever we want and not have to worry about answering for it, right? Yes, that's right. You can try to come up with some pathetic argument for materialistic morality, but it will fail on all counts.

Oh, wait. I just remembered a book I read a number of years ago by a dyed-in-the-wool Chuckie Dee Club member. He used the "evolution" of the Corvette as "proof" that things change with time. It didn't occur to this genius that Corvettes had a creator and designer (I actually have proof because I've been to their factory in Kentucky), the same way it doesn't occur to others that if man "creates life" or supposedly "controls" evolution, he'll just be duplicating what's already been done. Show me a Corvette that arose from the dust and changed itself with no one directing the change, and then we'll talk. I couldn't believe this guy would come up with such a DUMB example to "prove" evolution. He probably got a 'Vette out of it for giving them free publicity. All I'm gonna get from my book is a bunch of headaches. Apparently no one ever popped the question to him:

How did nothing become something and turn itself into everything?

CHAPTER XIX

Don't Know Why? Invent a Story!

The Corvette blarney was almost as good as Carl Sagan's explanation for why we suddenly wake up sometimes just as we're about to fall asleep. Want to know why? Just read *The Demon Haunted World*. You see, before we were people, we slept in trees. And once in a while, we'd begin to fall off a branch, and suddenly we'd wake up! And there you have it -- yet another Chuckie Dee Club just-so story invented to explain the unexplainable. And who's going to argue? After all, Carl Sagan was yet another high priest of Evolution (though I admit I liked the guy and considered him thought-provoking at least, while someone like Dawkins just has a chip on his shoulder).

Chuckie Dee Club Secret #12: ANY INVENTED STORY TO FIT THE FACTS IS ACCEPTABLE, AS LONG AS IT DOESN'T VIOLATE CLUB RULES BY HINTING OF DESIGN OR INTELLIGENCE.

Of course, people want to know "why" we do certain things, or "why" certain things happen. Well, if you are an Evolutionist, all you have to do is invent some story that "seems" to answer the question, and there you have it! Doesn't matter if it's TRUE. Doesn't matter if it can't be TESTED. Doesn't matter if it's even LOGICAL. As long as it "seems" to answer the question and has an interesting fictional story line, evolutionists are satisfied. Until another Clubber invents another story and they like that one better. As I said, there are currently over 100 stories about how the dinosaurs disappeared. STORIES.

That's how something called "evolutionary psychology" got invented, too, which is nothing but inventing stories to describe the roots of certain behaviors, and the biggest farce since Darwin wrote his version of The Good Book, which also is a racist book, but most people don't realize that. The full title of what most people know as *The Origin of Species* is actually *On the origin of species by means of natural selection, or, The preservation of favoured races in the struggle for life.* It's that "favoured" races part that most people miss. Of course, Chuck Darwin belonged to one of those "favoured" races as did most of his initial following, but some of the people he observed in his travels apparently did not. Chuckie Deeists will fight over that one and say Darwin wasn't referring to human races. That's because they haven't read his follow-up book on *The Descent of Man.*

I once taught a course on evolution's fallacies to a school that was attended mainly by African-American students. They were fully aware that if you took Chuckie Dee's theory to its logical end, then they were not quite as "evolved" as the white man, and they openly refuted evolution in part for that reason, and rightly so. Darwinism gives humanity yet another excuse for racism and looking down on another individual as "less evolved" or puffing oneself up as "more advanced." If you don't believe me, someone I know once boasted of being more highly evolved because he lacked chest hair. He was not joking. That's just one small example of how people can be influenced by this idiocy.

Which brings to mind another argument I love to use on Darwin Clubbers. If I believe in God, does that make me *more* advanced than people who do not believe in God, or, as Clubbers would have it, less advanced? I believe it means I'm

MORE advanced! Why? Because God is an abstract concept, and so my brain must be further evolved to be able to handle such a concept. So, HA HA! to the Clubbers on that one!! They have no argument. Because if all our brains are is a bunch of evolved chemicals, WHO ARE YOU to say that my beliefs are NOT more advanced than yours? How would you know that, and how could you prove it? You CAN'T, so don't bother trying.

We have now established that I am probably more advanced evolutionistically speaking than Chuck Darwin Evolution Club members, many of whom do not believe in God and, like Richard Dawkins, spend an awful lot of time fighting against something they don't even believe exists. Supposedly don't, that is. They're making a good bit of money off that fight, though I'm sure that has nothing to do with it (cough cough). Of course, that would only matter if it was a "preacher" making the money, right?

Oh, and we don't want to forget about "intolerance." That's another buzz word CDEC members like to throw around. For instance, if you say, evolutionistically speaking, that homosexuality makes no sense if we're trying to propagate the species, then you're INTOLERANT. Ok, so we need to deal with this brain-is-evolved-chemicals issue again. How do you define "intolerant" if you have nothing to define it by? I mean, if I don't want to eat rotten food, does that make me intolerant? Flies eat rotten food, so that means they are tolerant, but I am not. Correct? How in the world, if all you are is a fortunate mix of chemicals, do you come up with laws, morals and ethics that can be applied to all? Which evolved human decides that one thing is right and another is not? You have NO WAY of determining what's right and what's wrong and that's why we live in a more confused, rudderless, aimless, meaning-seeking world than ever. Argue with me all you want on that, but if you are an Evolutionist, you are now

BACKED INTO A CORNER. CHECKMATE, mate. There IS no right and wrong in Evolution. It's all in the way the chemicals turn out. So if one animal kills to survive, WHO ARE YOU to say humans can't do the same to their fellow humans if they want to? Who decides when it's murder, or when it's just killing for the advancement of evolution? It's a well-known fact that the last hundred plus years (since Darwinism) have seen more human death, injustice and suffering as a result of anti-God, atheistic, Evolution-inspired government polices than the rest of history combined, including every so-called "religious" war there ever was.

Take another example. WHO ARE YOU to decide what is "proper" marriage and what is not. Our ape ancestors invented marriage for some reason or other and got us all in a fix so we're fighting over it now. Let a man marry a man; let a woman marry a woman. Let five men marry one woman, or a half dozen women marry one man. Let kids get married, or adults marry kids. Which of our ape ancestors determined morality? Let a human marry his or her pet for that matter since we're all biologically related and connected by descent anyhow. Why should any of this concern us if we're nothing but evolved pond scum? I've had homosexuals get furious at me for bringing up this and similar arguments. Can someone explain why?

Come to think of it, I have not seen any ape tribunals lately. Do apes elect judges, and have lawyers and all that? Since they don't have lawyers, does that make them more *advanced* than we are? As I noted earlier, cats don't sit around contemplating the universe, and trying to figure out who they are and where they came from and where they're going or what's moral or not. No, but PEOPLE DO. Why? Who cares? If you're a Darwin Clubber, all you have to believe is that you came from slime, can live a slimy life if you want to, and you'll return to slime. No big deal, right?

CHAPTER XX

Proud to Be An Apeman

What about this idea that we're just a paw print above the apes, and there's really not much that separates us? After all, we share 95% or more of our DNA with apes, right? Well, it has also been demonstrated that we share about half our genes with bananas, but facts like that don't mean much to Clubbers. They don't like you showing them examples like this: GOD IS NOW HERE and GOD IS NOWHERE share ONE HUNDRED PERCENT of the same letters, but have completely OPPOSITE meanings. Sharing 95% of something is meaningless, unless you have all the facts. And the facts are that sharing 95% of our genes with an ape means absolutely NOTHING because it's the interpretation and expression of that genetic material that makes us who we are, and apes who they are. It's more evolutionary stupidity to try to prove their point about our "relation" to simian ancestors which, when all the facts are weighed, means nothing.

Just look at a few of the ways we're different from apes (or chimps, or whatever simian you want to choose as an Evolutionist with a chip on your shoulder). I am going to be as sarcastic as possible here, as we follow the typical Apeman in his daily routine from dawn to dusk. Here goes:

Mr. Apeman wakes up to the call of the alarm clock (which other apes invented). He gets up out of his warm, cozy bed, which has to have sheets and blankets on a comfortable mattress, and be made of nice (Norwegian?) wood, turned on a lathe, stained, and polished. Apeman knows he doesn't have much time, so he shaves and takes a quick shower, then goes and makes himself a cup of coffee. Outside, he'll find his morning newspaper, with news about all the other Ape people all over the world. He likes being informed.

Mr. Apeman then gets dressed smartly and is sure to put on deodorant, as he's not married yet, and there's an Apewoman at the office whose attention he's trying to attract. Apeman gets in his car, which was, of course, invented by Apepeople much more intelligent than he, and drives to work. Before he gets there, his car hits another Apeman's car, and they have to exchange pleasantries, and insurance information. As we know, all Apepeople carry insurance.

Mr. Apeman goes to an office, in an office building, in a city that was built by very smart Apemen. There he sits at his computer, and begins his day, just like all the other Apepeople in the city. He looks out the window and sees airplanes, helicopters, trains, cranes, cars, buses, Apepeople wearing all sorts of clothes, street sweepers, Apepeople carrying food they're eating, or briefcases, all sorts of Apepeople activities going on. Then Mr. Apeman decides he wants to listen to some music, so he flips on a radio, invented by other Apepeople. He enjoys classical music, which requires a whole group of Apepeople to play together in harmony. Eventually Mr. Apeman gets hungry. He decides to ask Apewoman out to lunch and they go to a nice restaurant, just like lots of Apepeople do every day, and sit in chairs, at a table, which is covered with a cloth, and has a candle in the middle. They also have glasses, plates, and silverware, and even napkins. Apepeople like to keep things clean and show some class, you see.

Apewoman has also dressed nicely. She was sure to comb her hair, brush her teeth, and put on makeup, as Apewomen do.

Mr. Apeman likes the way she looks. He's not just thinking that they need to have kids to advance Evolution, but that he'd like to live in the same house with her, and share his life with her, and Apewoman is thinking the same. So he asks her to marry him, she says yes, and they start planning a wedding.

Is this getting ridiculous? YOU BET. However, if you are still not aware of how different human beings are from apes, I'll be happy to continue. It is just plain STUPID to compare human beings with any supposed simian ancestors. Chuck Club members KNOW that; they just can't admit it.

And that's not the end of the stupidity. How many love songs, poems, ballads, and so on, have you read from the pen of a monkey? And just why do humans have to waste time on that stuff? Grab you a man or woman, have babies, and move on! "Git 'er done!" as we say out West. Why all the courtship stuff, lovey-dovey candlelight dinners, gifts, rings, wedding banquets, and on and on? How did *that* evolve? And WHY? Things were procreating and getting along just fine *without* all that stuff. Why did *we* need to add it? Oh, you can bet the Chuckie Club will invent some answer (after consulting with their Evolutionary Psychologists). They HAVE to invent an answer, because if something exists or is done a certain way, we want to know why, so they have to invent a story to explain it.

I know, I know. Apes can actually be Artists, and can also COMMUNICATE! Well, I've seen my share of what is nowadays called "art" and it appears to me that some human artists are doing a better job of imitating their ape ancestors than vice versa. Know any apes who could paint the Sistine Chapel? Would you pay them to do so?

Speaking of artists, one thing that always gives me a good chuckle is when cave drawings are found and some DEEP religious or ritual significance is assigned to them. How do you know they weren't just the doodlings of some bored kid who was stuck inside the cave because of the weather, while his "cave dad" was out hunting or inventing the wheel, and his "cave mom" was trying to get a fire going in hopes dad would bring home some brontosaurus steaks? No, there always has to be some deep spiritual meaning to the drawings.

As far as "communicating," when I see an ape reading this book and then explaining to me what it's all about, then maybe you'll convince me. Apes pushing buttons for "yes" or "no" is not very impressive, especially if you have to FEED them every time they get the answer right. I know some kids who would be starving if their teachers used that method with them.

But monkeys and other animals do use tools, right? Let's take another look at that one.

Chuckie Dee Club Secret #13: IF YOU CAN'T CHANGE THE MEANING, THEN BROADEN THE MEANING.

Thus, if by "tool" most people understand a hammer, or a screwdriver, or a cordless drill, evolutionists simply include things like sticks and stones under the definition of "tool,", and another evolution problem is solved. That's not to say that a stick or stone can't be used LIKE a tool, as we've already discussed. But if you pick up a hammer or a screwdriver, is there any question in your mind that it's a tool? How about if you pick up a rock or a stick. Do you say to yourself, "Hmm, it's a TOOL!" Or do you say, "Hmm, it's a rock (or a stick)!"

Here's another example. Lacking a gun (a real tool) with which to shoot me, some of you would be satisfied with stoning me instead. You would be using a stone LIKE a tool.

But a stone is NOT a tool, in the everyday understanding of the word. If an otter uses a rock to break an oyster shell, is it using a "tool" or is it just using a rock to break an oyster shell, and COULDN'T CARE LESS what it's called or whether it's a tool? Then why do you need to concern yourself with what the otter is doing? Just leave it alone!

Why do we humans always have to attach some human behavioral meaning to the actions of animals? Sure, in some cases, it may seem logical, like with courtship displays that lead to having babies. But if two male animals are playing around, all of a sudden they're exhibiting "homosexual behavior"? Or is it that we're just seeing what we WANT to see in their behavior, to try to justify our own? Homosexual behavior, from a strictly evolutionary standpoint, makes absolutely no sense.

First, why did the goddesses Mother Nature and Time invent two different sexes if we could procreate without them? Second, how does homosexual behavior serve to help the human race survive and continue to move on into the future if two people of the same sex can't procreate? Homosexual behavior conveys no "survival advantage." Why don't we hear Evolutionists arguing these points? Not politically correct? Not popular? Can't invent any stories to cover it over? Why didn't the goddesses Mother Nature and Time just make us hermaphrodites? Then we wouldn't need each other at all, and we humans would have been spared all this arguing and societal conflict about the issue.

On the other hand, if we're just evolved chemicals, then who's to say any sexual behavior is right or wrong? Why not have sex with the apes down at the zoo if you feel the urge? And swap wives or husbands or have multiple partners or feed your daughters to the raging teen hormone lions. If you're an Evolutionist, and you REALLY believe what you SAY you believe, and you argue that certain sexual behaviors that YOU disapprove of are somehow "wrong," then you are nothing less than a hypocrite, because you're not practicing what you preach.

What about the "survival" argument. If you believe Chuckie Deeist fanatics of the world, Mother Nature "invented" clam shells, for one example, to protect clams. Well, WHAT THE HECK WERE THEY DOING FOR PROTECTION BEFORE THEY HAD SHELLS? And now that otters have figured out how to break the shells, alotta good all that evolving did! How come clams aren't evolving ever harder shells to foil the otters? The whole "survival of the fittest" argument is a tautology and easily refutable. Our earlier discussion about bacteria versus dinosaurs is just one example.

Let's play, "Invent the Evolutionary reason for everything." What you're saying is that the *purpose* of a shell seems *obvious*, but HOW DID EVOLUTION KNOW THAT? Where's your story for why the otters got around the survival myth by learning to break clam shells and the clams aren't responding? Where are all the failed shell experiments in the fossil record? There AREN'T any. Not a one. Clams are clams, shrimp are shrimp, crabs are crabs. I've seen fossilized crabs, and they look just like the one I had for dinner a few weeks ago. I don't look at a crab fossil and say, "Gee, it looks like a blob that's evolving into a crab." No, I look at it and say, "That's a CRAB!" And that's the same thing Chuckie Dee Clubbers do, but they imagine that it WAS a blob at one time, with no reason to do so. It's all in their imagination. But let's give the fossil one of those fancy Latin names, like *Crabilus habilus*, so no one will think of the fossil as an actual crab that is no different from a modern crab and not millions of years old, but rather they'll see it as some mysterious ancient life form that just happens to look like a crab. I'm just tired of people being duped by this stuff.

You know, if we're not eating our ancestors, or shooting them, or making clothes out of them, we're putting them in zoos. At the London Zoo a few years ago there was a cage in which humans were cavorting around like they were any other animals. This is taking Evolutionary stupidity almost to its maximum level. I say "almost" because I'm sure someone will one-up the London Zoo sooner or later. But at least the zoo people were not being hypocrites, and were living up to (or down from might be better) their beliefs. Except that the human zoo specimens were clothed, and smiled at the zoo patrons, who were much happier to have paid to see these nice, "favoured race" folks behind bars, than to go to a *prison* and pay to see the folks *there* who are behind bars, who would likely not be quite as friendly, as they're not being paid to do what they do.

Speaking of which, I almost forgot the shrew stupidity. You know, that little mousy thing with a cute nose that moves around like a tiny elephant's trunk? Well that was your MAMA, and I'll bet you didn't even know it. If you did, and you believe it, then shame on you.

Yes, the lowly little shrew evolved into all the land mammals we've ever known, according to Darwinists. Don't ask why we still have shrews around, and why they all didn't keep evolving, and why they're not evolving now. The fact that not ALL things evolved, but some stayed just like they were millions of years ago is one of the conveniences of Evolution. OBVIOUSLY some shrews became other mammals and then people, and OBVIOUSLY some stayed just like they were -- AS SHREWS, DUMMY! Can't you see that some of them turned into people and some didn't? You can't? Well, then we need to indoctrinate you more. If you don't believe my shrew story is true, look it up. If you believe it's STUPID, congratulations.

CHAPTER XXI

Not in *MY* School, You Don't!

Brainwashing and Political Correctness are definitely two of the hallmarks of membership in the Chuck Darwin Evolution Club. Do you DARE to question Darwin in the classroom? I think not! The little demons of Darwinian darkness will come after you with their pitchforks. You can question God. God is big enough to handle it, I suppose. But question DARWIN? We don't do such things. Not in our schools, and certainly not in our universities. We're too SMART to question Evodelusionism. We *know* Evolution is a *fact*, so why do we need to question it, right? If something in the Evolution myth doesn't make sense to you, just take it by faith that you don't have enough education to understand it, or that "someday" the answer will be there.

"Gould said it. I believe it. That does it."

If you upset the boat by asking too many questions, your teachers, who usually don't know how to answer your questions about Evolution anyhow and need a quick "out," will either brand you as one of those pesky "creationists" or IDists (Intelligent Design believers), or say something like "you're not a scientist" and that's why you don't understand. What they can't see, and what they can't admit, is that they too have been brainwashed into believing what they believe. And for the brainwashed, it's easier to NOT ask tough questions or seek answers. After all, if you're comfortable, why bother having your beliefs threatened? You don't need to find out that yet one more thing you believe isn't true. Is anything sacred any more? Sure! DARWINISM IS!

And don't dare acknowledge your doubts in public, as more and more scientists and others are doing. You'll be branded by the Chuckie Dee PC squad before you can say "punctuated equilibrium." After all, they have their Religion to protect. Not to mention their jobs and reputations among peers. And only *stupid* people say that Evolution is stupid, right? Wait a minute... let me think about that...

DON'T LET THEM INTIMIDATE YOU!

You KNOW that evolution can't be true. You *know* it. Come out and admit it! As I've said before, I don't care what else you believe, but if you really believe Evolution, I can only conclude that you have not *thought* about it at all. If you HAVE thought about it, then there is no way you can really believe in your heart that it's true, and it's time to just come out and admit it. If you believe Evolution, you're believing (and perhaps even defending) something that you KNOW is not true, and you know CAN'T be true. So stop being a coward, and come out and say it.

You can send me your thoughts and opinions at Feedback@EvolutionIsStupid.com. It is my sincere hope that eventually many who are sitting on the fence, waiting for others to make the first step, will get down off the fence on the side of truth, and simply admit that Evolution is bankrupt. Many of you who have escaped from the cult of Evolutionism have stories to be told. Go tell it on the sea-shell-covered mountains and everywhere!

So.......

How did nothing become something and turn itself into everything?

It didn't.

Feedback Responses

One of the criticisms I heard about the original version of this book is that it was not "scientific." Well, guess what? It was never intended to be a science book. But I got to thinking that maybe it would not be a bad idea to expand the original and append a section that covered some of the issues I've had to address over the years, for those who might be interested in more in-depth discussion, or just those laypersons who wish to deepen their knowledge of why evolution is stupid and an untenable theory. Now of course I realize that for some others the following still isn't going to be "scientific enough," but you know, you just can't make some people happy. I even tried to throw in a few real scientific-sounding terms to appease them, but doubt it will work.

The following discussions are taken from responses to the content of this book both online and in person, from around the world. No names are given and some of the content has been edited or abridged for clarity and brevity, or embellished where further detail may have been called for. Topics cover a wide range of issues (and I do sometimes repeat information, as that's the best way to learn it) and I believe that the reader will find them entertaining, enlightening and informative. I did try to maintain the occasional sarcasm that characterizes the rest of the book, and I hope it makes you mad if necessary so you'll go find out for yourself whether what I've said is true or not.

It is interesting how often writers and critics attempt to focus on religion, or God, or the Bible, or otherwise turn the conversation away from the focus on Evolution. It is also interesting how many hostile responses I've received over the years. Those for the most part have not been included as usually I'd just ignore them and not respond, especially if the writer did not put a name behind his or her words.

In case it's not obvious, the original feedback is denoted in bold by "**FB:**" and my responses follow that. Enjoy, and I hope we all continue to learn.

****

FB: How did God create the Earth?

My desire is to keep the focus on evolution itself, so I'd rather not get into issues of God or creation for now. Just evolution. So, now, may I ask you how evolution created the Earth, and where your proof lies? [I did not receive a response to that question.]

FB: When we can't answer a question, we resort to "God did it." The "god of the gaps." So is God some sort of magician?

No more so than evolution, which seems to always come up with just the right evolutionary path for each situation. If one evolution story doesn't work, all you have to do is invent another one that sounds more plausible to the unthinking mind. And if you just don't know the answer, you simply have faith that the science god will come up with magic answers sooner or later. The god of evolutionary gaps!

FB: I believe there is no god because there`s no evidence for God. I can't say it's a fact, but I just don't see evidence that there is a God.

Exactly what evidence would you expect to see? [I did not receive a response to that question, but usually the response focuses on the evil in the world and how God could "allow" it.]

FB: You do not understand evolution. It has nothing to do with the Big Bang or cosmology. Evolution works on living things and is about survival and change in living things over time.

Evolutionary cosmology begins with the Big Bang, without which there would be no Earth on which biological evolution could occur.

Evolution does not "work" on anything. You are confusing "change" and "adaptation" both of which work with ALREADY EXISTING genetic information and physical capabilities, with classical evolution from "simple" to complex, which simply does not occur, and never has. Evolution of life from inorganic matter is impossible, no matter how much Time you add. Even if you begin with a gene (we'll take it for granted that it just magically appeared over time, and was able to reproduce itself), in order for classical evolution to occur, the genome must increase in information content, and that simply does not happen (mutations sometimes duplicate information (which, again, is ALREADY present), but mostly corrupt it). And even if it did, you still need a mechanism in place to express that information and put it to use. There is no such mechanism that could have started it all and increased information and complexity without an intelligent source.

In other words, Chuck Darwinists tout things like mutations and natural selection, but both mutations and selection work with genetic material that ALREADY EXISTS, and do not create anything new. Nor do they explain where genetic information and genes themselves came from in the first place.

As far as "survival," if you think about it, there's a whole lot more to life than just "survival" and the whole survival of the fittest thing is a tautology. As I point out in this book, bacteria were surviving just fine, and there is absolutely no reason, apart from evolutionary mythology and storytelling, that they ever had to change into anything else just to survive. There is CERTAINLY no reason they had to become more complex. It's just evolutionary story weaving.

Obviously, if something is extinct, we're going to say right away that the reason is that it was not "fit" so it did not survive. But that's just the evolution myth. There were actually many MORE living species - plants and animals - long ago than there are now. So really it's not that what's alive now was any more "fit" than they were, but rather that the great diversity of life that once existed is slowly disappearing. Within that great diversity was also genetic diversity, hence animals and plants could diversify, spread out and occupy ecological niches. But as they did so, they actually became LESS able to survive OUTSIDE OF those niches, and that's why they became extinct. So, what really happened was not that plants and animals, thanks to evolution, were BETTER able to survive but rather, because of adaptation and genetic

DETERIORATION, they became LESS able to survive except under special conditions.

For example, you start with an original bear. That bear contained the genetic information to diversify into various kinds of bears, which lost some of the genetic information the original bear carried, thus rendering them able to best survive in special conditions, such as polar bears, which survive best in arctic climates.

Antibiotic resistance is another alleged "proof of evolution" that Chuckie Dees like to throw in front of those who don't know any better, but it has NOTHING to do with evolution. Resistant bacteria already exist in the environment, and when an antibiotic is introduced it kills off the hardy (normal) bacteria, leaving only resistant ones, which then thrive in the presence of the antibiotic. The resistant ones are in reality LESS able to survive, because if you then remove the antibiotic, the resistant strain dies off, and the hardy strain takes over once again because it does not need the presence of the antibiotic to survive like the resistant ones did.

FB: Without an understanding of evolution, medical science would not advance, and cures for things like cancer would never be found.

Evolution has NOTHING to do with finding a cure for cancer, or any other scientific or medical discipline. If a cure is found, it will be because intelligent people worked with their intelligent minds to find one, using real science and medicine, and not because of some blarney explanation of how everything came to be.

There is simply, plainly, no scientific or other discipline that needs the theory of evolution in order to advance. None.

FB: Redshift is another proof that the Big Bang is true. Do you also believe the Earth is flat?

I am an amateur astronomer, and I have found that what you are *not* told in the average astronomy text is that there are a few other JUST AS VALID explanations for redshift besides that the universe is expanding from a Big Bang. Among them are tangential motion, gravitation, and a simple loss of energy over time and distance. I find it unimaginable that a beam of light could travel for billions of years from its source, and still have the same energy that it had when it left that source.

As for people (especially Bible believers) thinking the world was flat, that is an old canard, as the Greeks knew well that it was not true, and the Hebrew Bible states that the world "hangs upon nothing" in the book of Job and speaks of the "circle [also can be translated 'round' or 'sphere'] of the Earth" in the book of Isaiah. If people believed it was flat, Columbus would not have undertaken his journey, which followed countless other voyages by ship before his time, by experienced mariners who no doubt were intelligent enough to realize that horizons don't just disappear for no reason.

The flat Earth idea comes in part from a semi-fictitious account of Columbus's life and voyages written by Washington Irving, and not from factual history.

This might be a good time to point out that the majority of (or perhaps all) people who believe in things like astrology, psychics, the paranormal, flying saucers, extraterrestrials, trance channeling, crystal powers, auras, ghosts, reincarnation, and so on, without *any* scientific evidence to back them up, are also evolutionists. Carl Sagan, one of the biggest proponents of the non-scientific false myth of evolution, decried all the weird things people believe in his book *The Demon Haunted World*, a classic case of being able to clearly see the faults of others while ignoring his own.

FB: You are a close-minded bigot who refuses to see the facts.

Actually I have no more open mind than you do. Because I WAS an evolutionist at one time, but now I'm not. When I was an evolutionist, I really didn't know why, but now that I'm not an evolutionist, I know EXACTLY why, and that is because there is not a shred of evidence in favor of it. I believed evolution blindly, just as you do now. The 'You don't have an open mind' accusation is yet another smoke and mirrors ploy that evolutionists use to get out of having to defend their indefensible position. Please don't fall into that trap.

FB: What about the appendix and other vestigial organs? Are they not proofs of evolution that you ignore?

You have not researched the appendix very well. Nor other so-called 'vestigial' organs. Just because you are not aware of the function, or former function, of something does not mean it does not, or did not, *have* a function. There were once well over 100 so-called 'vestigial' organs, most of which have since been clearly determined to have a function. So you see, in your mind the disjointed thinking is:

"I don't know a function for the appendix, so that PROVES that particles turned into people all by themselves."

You're using your non-knowledge of the function of an organ to justify your belief in the myth. The fact that the appendix exists doesn't explain where it came from to begin with. Vestigial organs "prove" nothing.

FB: You're also ignoring the fossil record, which abundantly proves the progression of evolution from simple to more complex.

Show me this fossil record that has convinced you that evolution is the reason we're here today. I want to see a clear progression from particles right on up to people, not some imaginary tree that an artist invented. I want cold, hard fossils, and I don't want just bones and shells put next to each other in some imaginary progression. I want to see all the soft parts, too, and how they evolved.

Then I want to see all the failed evolutionary experiments. I don't know of a single fossil where evolutionary scientists have not applied some purpose to an appendage or organ, so where are all the useless examples of evolutionary "tinkering?" I want to see all the useless intermediate stages of development.

I'm a real doubting Thomas, and I won't be convinced until I see all the stages that led to the human nervous system, the human digestive system, the human excretory system, the circulatory and lymphatic systems, and on and on. I don't want some evolutionist's imagination. I want to see all the stages that led to these things right there in the fossil record. And more than that, I want to see how all those systems managed to come together in one creature, and function together so beautifully. Don't give me your evolutionary faith belief story. I want FACTS and EVIDENCE, ok? Not just a few bones in an imaginary progression.

FB: Varves are another proof that long geologic ages are real.

Varves prove nothing other than that sediment bearing water produces them. They can form rapidly and do not need millions of years, as was demonstrated when layers of sediment were formed in a few hours in the Mount Saint Helens eruption here in the States. Varves don't need the right TIME to form (there's your goddess Time again), they need the right CONDITIONS, just like fossils. Further, here we go again with the typical evolutionist *non sequiturs*. How does the presence of varves prove that particles turned into plants, animals, and people all by themselves? Here we go back to my bike trip across America and how it proves that grizzly bears evolved in Alaska.

FB: If there was a global flood, where did all the water go?

Well, tell me what happened to all the water from the flood on Mars, which evolutionary scientists insist occurred. If Mars could have a flood, and all the water disappear, is there some reason it couldn't happen here too? [Note: There are mechanisms that explain where the water went, but we'll not get into that discussion here.]

FB: You do not understand how natural selection works. It is a cumulative process whereby favorable adaptations accumulate into a more and more complex organism.

Could you please give an example of what you mean by "cumulative" natural selection? Exactly what evidence do you have that something "accumulated" apart from the fruits of someone's imagination? The fact is that natural selection produces absolutely nothing new. That is a KNOWN FACT. Natural selection works with what already exists, and usually results in a DECREASED ability to survive outside a specific environment or niche.

For example, fish that live in caves can lose their eyesight, which may render them better able to survive as they turn to other senses to feel their way around (just like humans do). So this is touted as an example of natural selection and evolution in action. However, a LOSS of eyesight does not answer the question of where eyesight came from to begin with, which, by the way, requires not just eyes, but an ability to use light, transmit it to a brain, and the ability of the brain to interpret it. So any imagined "sequence" of eye evolution has to take into account not just the structure of the eye, but musculature, a circulatory system to nourish it and keep it working, a brain to interpret and control it, and so on.

Again, natural selection works with what ALREADY exists, and does not "create" anything new at all. Further, assuming I understand what you mean by "cumulative natural selection," you have to REALLY stretch your faith to believe that this sort of "cumulative" mechanism resulted in EVERY KNOWN LIVING THING, when there is NO evidence for any such "ladder of life" in the fossil or any other record. There should be COUNTLESS FAILED "EXPERIMENTS" if what you're saying is true, and there simply are not. There are many examples, such as the blood clotting mechanism or caterpillar metamorphosis into butterflies, that could not possibly have resulted from numerous accumulations. Darwin said that if it could be demonstrated that something did not result from numerous accumulated minor changes, his theory would fail. It has failed time and again, but the fanatics are not about to give it up.

FB: There seems to be a moratorium on ape evolution, as apes do not appear to be becoming humans any more. Nor have I seen something crawling out of water trying to evolve. Why is that?

Yes, it appears there is a moratorium on ape evolution, as I have not seen, nor has anyone else, any apes showing the least desire to become humans. I have, however, seen a number of humans regressing some toward apehood.

As for "something crawling out of water," let's go back to the question of why they'd ever have done that to begin with. We'd have to attribute some mysterious (shall we say "metaphysical"?) force to matter that urges it to become more complex. There is no reason in creation why it should do so without some outside impetus, and there is no impetus unless you believe in some imaginary evolutionary "force" that moves things toward higher complexity. That is the reason the "anthropic principle" was invented, in fact.

So it is preposterous to believe that fish moved onto land and turned into flamingoes, felines, and Freddie. They had no reason to do so. They still have no reason to do so, which is why we don't see that happening. I have, however, seen a number of humans regressing toward fishhood. The evidence can be found on beaches all over the world in summertime.

FB: Experiments have shown that the early atmosphere of the Earth could have had molecules in it that became the building blocks of life.

I discussed the molecules that 'could' [as you say] have been in the atmosphere a long time ago in this book. That is the Miller-Urey experiment which has been INVALIDATED on the following premises: First, no one can prove what any "original atmosphere" was made of. It's all guesswork. The alleged atmosphere that Miller used has now been challenged by new information. He and Urey proposed a reducing atmosphere, and now scientists (as often happens) have taken an about face and said it was probably oxidizing.

Second, putting chemicals in a flask requires a creator and designer, who creates and designs the experiment, sets all the parameters so that at best he gets the outcome he's hoping for, starts the experiment, and ends it. The experiment did not create and run itself. It was created and run by an outside intelligence. That alone negates it as any sort of "proof" of evolution. I make that clear in this book. If Miller had not ended the experiment when he was satisfied that he had some results, whatever was in the flask would have eventually been destroyed.

(I do recall having left something in the oven once or twice, and though it formed a gloppy molecular structure at first, it was eventually destroyed completely, and a mess to clean up, to boot! But, of course, if Miller and Urey's experiment were still sparking away, there'd probably be a human being in that flask by now, wouldn't there?)

And now let me throw a monkey wrench (pun intended) into the equation. If you know anything about chemistry, you also know that energy can BREAK chemical bonds, which is EXACTLY what would have happened to any so-called complex molecules that formed in the original alleged primordial 'soup.' The bonds would have been broken by UV rays and cosmic rays. Also, the molecules would have dissipated in the medium (ocean, pond, whatever) in which they formed. So once again your evolutionary religious faith fails when put to the test.

Finally, the fact that a few amino acids were CREATED by an experimenter means absolutely nothing. Hence my diatribe against the whole "building blocks" scenario, which is meaningless if there's no intelligence to do something with the "blocks." And hence, another invalidation of Darwinism.

It's the PHILOSOPHICAL CONCLUSION they and others drew (and continue to draw) from it that's at issue here. Any scientist can spark a bunch of chemicals and make more complex ones from them. But to then say "VOILA'! HERE'S WHERE LIFE CAME FROM!" is nothing short of arrogant imagination.

FB: Changes in DNA are what drives evolution.

The fact that something happens to DNA has NOTHING TO DO WITH EXPLAINING WHERE IT CAME FROM IN THE FIRST PLACE! Sorry for yelling, but I tried so hard to get that point across in this book. Evolutionists point to mutations in DNA, copies of DNA, substitutions in DNA and proudly trumpet their "proofs" for how evolution took place because of changes in DNA. They conveniently avoid telling us HOW DNA EVOLVED in the first place. DNA has incredible amounts of information stored in its molecules, and here we are saying that all came about on its own? Well, it is a well-known SCIENTIFIC FACT that information does not arise on its own. It has to have a source. It has to have a code. It has to have a mode of transmission, a mode of reception, a mode of translation, and apparati in place to utilize it. This does not happen on its own. Thus, another invalidation of Chuckism.

For example, if I say to you, "paint it red," my brain is the source of the information and encoding of the letters in the words "paint it red" (as I already have an understanding of what I wish to convey), my voice is the mode of transmission, your ears are the receptors, your brain the instrument of interpretation, and then you can apply it by using your hands and arms and a paintbrush.

If I say those same words to someone who does not speak English, they are meaningless. So you see, for DNA to arise on its own until it contained the billions of bits of information necessary to make a living thing, and to transmit them, translate them, and make something out of them, with NO direction or "interpreter" whatsoever, is a miraculous act far beyond anything anyone could ever imagine, and is science fiction, not fact.

FB: What are your sources? Can you provide sources and documentation for what you claim to be true?

My first 'source' is my brain - I THINK about things and QUESTION them, rather than just believing what someone tells me.

'Sources' are not going to matter much. I've got plenty of 'em. If you're going to choose to believe in evolution, my 'sources' aren't going to change a thing. Let me explain.

You see, if I quoted a secular source in support of my beliefs, you'd say I misquoted or misunderstood it. If I quote, say, a "religious scientist" you'd say he wasn't a "real" scientist. If I cite a particular book in my defense, you'll just cite another in yours. Bottom line: We've both made a choice, and you're going to defend yours, and I'm going to defend mine. It remains for the observer to decide which side he'll take once the facts have been presented and the logic and reason defended.

One thing I've found interesting in many years of dealing with evolutionists is that they RARELY require documentation of anything an evolutionist tells them. For instance, our friend above obviously had no problem believing that Stanley Miller and Harold Urey knew EXACTLY what the atmosphere of the earth was composed of BILLIONS of years ago. After all, they're "experts" and who is to question their assumptions? But if you DO question them, you'd better provide documentation for why you disagree.

FB: Once again you've demonstrated a lack of knowledge about the process and theory of evolution. You should be ashamed of yourself. Evolution is a tinkerer. When something doesn't work, it's discarded. Small changes accumulate over time to become large ones.

Let me begin by pointing out that YOU have a vested interest in defending evolution. I don't. You're a doctoral student in biology, right? So if you don't defend the myth that you are required to believe unquestioningly by the priests who are teaching you their religion, you'll be booted out. I don't have to be concerned about that. If I say evolution is a pile of doodoo, I won't lose my job or be refused my degree. You will because you're part of a religious system that expects unquestioning obedience and allegiance.

I always love it when evolutionists anthropomorphize. Can't avoid it, can you? As if Natural Selection has some ability to know what it's doing. That's why words like "tinker" and "create" and "made" are used so often in evolutionspeak. The fallacy of your statement should be patently clear to any thinking person, as it was even to Darwin. If what you say is true, then that "tinkerer" should have left BILLIONS and BILLIONS of FAILED and JUNKED experiments laying around in the fossil record. Is that what we find? Not even a chance. What we find are well-defined structures to which clear functions can be attributed, not a bunch of half-formed pieces that your "Natural Selection Tinkerer" discarded.

"Small changes" in what? IN WHAT? Let me answer: in what ALREADY EXISTS. Do you understand that? You need to tell me where the material in which those "small changes" are taking place came from IN THE FIRST PLACE. Then tell me what evidence you have that those "small changes" turned into large ones with time. NONE. They didn't. It's nothing but your imagination, and there is no mechanism known by which genetic information could have arisen on its own from base, inanimate material with no outside input, and then that information increased to make more and more complex organisms. INFORMATION JUST DOES NOT WORK THAT WAY.

Further, you then have to come up with some mythical explanation as to WHY THINGS GOT MORE COMPLEX. Is there some magical law at work that forces things to go from simple to complex? No, there is not. However, GOOD SCIENCE tells us that everything is tending toward entropy, or toward less complexity. And don't give me the line about crystal formation proving matter can organize itself into more complex things, or life itself going contrary to thermodynamics. The NET result is always a gain in entropy (disorder).

Sorry that I have a poor understanding of evolutionary theory. But at least you call it a theory and not a fact or law of science, which it clearly is not. "Ashamed" of myself? Why's that? If I'm only a bunch of random molecules with no meaning to my existence other than that which I invent to try to give meaning to it, what is "shame" that I should be "ashamed?" I'm only trying to explain my existence according to what my evolved brain molecules make of it, right? So, why is my explanation any more "shameful" than yours?

FB: [From same writer concerning the issue of Morality]:

I particularly like your ending, though. "Who is to say [that morality cannot exist without God]?" That is MY point exactly: Who are YOU to tell ME that I am wrong? Why should we listen to Mr. PhD candidate, or any other PhD? Does that make YOU God? I think so! You're saying, "LISTEN TO ME!!! LISTEN TO ME!!!!" and then asking the question, "Who is to say?" Apparently you believe YOU are to say, no??!! If what you believe is true, viz., that there is no design in nature, and that the only purpose in anything is what our evolved chemicals which we call a brain assign to it, then who are YOU to say that what you believe is the "right" thing, and what I believe is "wrong"?

You don't even seem to be able to comprehend your own illogic! WHY should anything that evolved be "moral?" WHAT is "morality" if, by evolution, it's nothing more than the product of a bunch of random molecular processes? You have NO BASIS for morality, other than what YOU decide it is. And if I decide differently, then that's too bad for you. My evolved molecules just think differently from yours.

FB: You have faith in what you believe. Science is about facts, not faith.

Everyone has faith in something. It's just that often it's faith without substance. The evolutionist is a person of tremendous faith! How can one say they believe in so many things that no one has ever seen without having faith? The big bang? FAITH! Transitions from particles to plants, panthers and people? FAITH! Planets forming from exploded star balls? FAITH! Brains to contemplate it all evolving on their own with no direction? You got it: FAITH! That's why a couple of guys wrote a book entitled, *I Don't Have Enough Faith to Be An Atheist*. Because the atheist/evolutionist has perhaps the most faith of all!

Here's another thing that absolutely amazes me about the evolutionist and his faith: His only hope is to return to the dirt and become fodder for future evolution, but he'll fight you to the death if you try to convince him that there's something more to life than his stupid theory. And if you try to convict him of the stupidity of it all by pointing out that if he's consistent then he has to admit that whatever he believes is nothing more than the outcome of some evolved chemical reactions and therefore meaningless, he'll do all he can to convince you that his belief has more meaning than yours! It's utter insanity!

The bottom line is, there are only two possibilities: a Creator did it, or blind chance Evolution did it. Both are faith beliefs, but believing something by faith does not have to mean believing it blindly. I believe when I sit in my chair it's not going to collapse. That's faith, but it's based on observable realities. Some day it might collapse, but that does not then negate my faith that the chair should not collapse when I sit in it. There's just another explanation for why it collapsed that explains it just fine.

FB: The information in RNA could be encapsulated for a period of time [while it was evolving into DNA.]

This means nothing, really. The information is already there. The RNA is already there. Encapsulated information? What good is it? Suppose I put a message in a bottle and send it out in the ocean and it winds up on a shore somewhere and is subsequently buried. What good is my 'encapsulated information'? However, if another human being finds it, who can open the bottle and read the message it contains, and make some sense of it, then the information becomes something useful, perhaps. But the whole process, from beginning to end, was directed by intelligence.

FB: The evidence for the Big Bang is overwhelming.

You are committing the classical evolutionist fallacy of confusing 'evidence' with 'interpretation.' The EVIDENCE may exist in the form of redshift, CMBR [Cosmic Microwave Background Radiation], etc. But your INTERPRETATION is that they support the Big

Bang. There are other interpretations of that evidence that are NOT consistent with Big Bang theory. I believe I've already presented some, such as the fact that the CMBR could be simply the temperature of interstellar space (after all, those stars put out a lot of heat, don't they?), and that redshift has other explanations such as tangential motion, or gravitational stretching, or simply light becoming "tired" (losing energy) after traveling great distances. But the evolutionist ignores the other interpretations because they conflict with his religious faith.

FB: Have you heard about the enzyme that evolved to allow nylon to be digested? How do you explain that apart from evolution? It is a clear example of new information that was not there before.

The nylonase [the digestive enzyme] example has been challenged of course, but apparently you either ignored or chose not to read the articles debunking it. Here's a great example of evolutionists jumping to their philosophical conclusion before all the facts are in. An enzyme is discovered that can digest nylon, a man-made substance, and suddenly the evolutionist jumps to the conclusion that this PROVES that particles turned themselves into people.

The fact that you have to latch onto something so trite and pathetic to prove your bigger picture shows just how desperate you are, not how much proof you have! If you wait till all the facts are in, the opposition's contention that the enzyme that digests nylon is not an example of new information at all, but rather a recombination or change in information that was already present, will prevail. And in fact, subsequent experimentation seems to be bearing that out. Do the research.

FB: If evolution is not true, do you have a better explanation? Evolution is a great explanatory theory.

Your argument amounts to the following: Evolution is demonstrably false, but since we don't have a substitute (or I should say we do, but don't want to accept it as such), we can't trash it.

Let's put it another way: We're going to choose to believe a lie because we haven't come up with a better explanation.

Or another way: This tool is totally useless for the job I'm trying to do, but I'm not gonna trash it till another tool comes along.

Or another way: I believe in flying saucers even though there's absolutely no proof they exist, but because I haven't heard a better explanation for those strange lights I saw the other night and that other people say they see now and then, I'm going to believe in them anyhow.

Or one more: I believe astrology is the best explanation for what guides our daily activities (though there's no scientific proof for it) because that's the best explanation we have for what's guiding us, and until someone comes up with a better one, I'm going to believe it, despite the fact there's not a *shred* of evidence that it's true.

Your just-so statement that evolution is a 'great explanatory theory' amounts to admitting that it's a great basis on which to build fairy tale stories that supposedly explain where everything came from. And we're gonna believe those fairy tales come hell or high water. Evolution is an 'explanatory' theory only in that its adherents are very adept at inventing explanations to keep the theory alive. You have the same facts that I have, and that thousands of other scientists who do NOT believe in evolution have, and we use different 'explanatory' theories to explain the facts. Once again, you're confusing the facts with the theory. Evolution explains nothing. People explain things. It's called interpretation, and it's a faith belief, and you are one of its fanatical defenders.

[You stated further that,] "The National Science Teachers Association (NSTA) strongly supports the position that evolution is a major unifying concept in science and should be included in the K-12 science education frameworks and curricula." Certainly it's a unifying concept! It unites those who would rather let other people do their thinking for them into a solid group of religious fanatics who will defend their faith despite all the evidence to the contrary.

If evolution is such a unifying concept, why do so many evolutionists disagree with each other about everything but the theory itself (which has to be preserved at all costs!), and why are we having this debate? Apparently you're unaware that thousands of scientists, and millions of laypersons worldwide don't even accept evolution as a valid explanation of origins. Check out the Dissent from Darwin website for starters.

FB: How can you say radiometric dating is based on assumptions?

There is a big difference between testable assumptions and non-testable ones, or assumptions that are deliberately designed to support one's viewpoint or thesis. For example, ALL radiometric dating is based on assumptions that are designed to give old ages, or pre-determined ages (carbon 14 is about the only somewhat valid method because it can be roughly calibrated by using objects of known age, but it can only be used best on items that fall within the timeframe of recent human history). If you remove the assumptions and just work with the bare facts, you quickly realize that the 'millions of years' assigned to rocks and fossils can't possibly be proven.

There is no way to know that no daughter product (e.g., lead from the decay of uranium) was present when the rock was formed. There is no way to know that decay rates have always been constant over millions and even billions of years. There is no way to know how much original parent substance (e.g., uranium) was in the rock when it formed. There is no way to prove that parent or daughter products were not affected by environment via leaching, or chemical processes, or heat, etc. It simply is impossible to demonstrate unequivocally that a rock is so many millions of years old.

Another issue is the assumptions made regarding fossils. You find a couple of different bones and build an entire belief system on them, drawing imaginary 'transitional' lines between them. Even Stephen J. Gould recognized that the average 'evolutionary tree' only shows fully formed creatures at the tips and nodes, but nothing in between. You can only deduce a few things about the actual creature just by its bones. That's why artists' renditions (speaking of assumptions) of so-called pre-humans often show completely different-looking creatures, based on the same set of bones (check out how Neanderthal Man has changed over the past few decades!).

FB: Evolution is only a theory and is not testable, like for instance Max Planck's quantum theory.

You are absolutely right. Another word, perhaps, is hypothesis, and a hypothesis needs to be testable. The assumptions made by Planck *et al* were, and are, testable ones. That's what REAL science is about. You cannot test whether a shrew turned into a human (which, in case you didn't know it, is one of the assumptions evolutionists make). You cannot test whether a rock really is 5 million years old. You cannot test whether an imaginary 'primordial soup' generated cells. Even if you could do that in a laboratory experiment, that would not prove it actually happened in nature, but rather just that 'maybe' it 'could' happen. It never will, of course, but I'm using an extreme example. You cannot test whether clouds of interstellar gas really do collapse and form stars and planets - intuition and direct observation seem to contradict that outright, in fact. You can only observe what you're able to, and then INTERPRET those observations according to your pre-determined paradigm. It's that pre-determined paradigm that needs to be tested, and the evolutionary one fails every time.

By the way, I love the fact that these geniuses can look at a statue of a man and see design and a designer, but can't see it in the man himself. Ya gotta love it. They'll tell you all about what such and such is used for or what its function is, but there ain't no design and purpose there, right? So WHY ARE THEY LOOKING FOR IT THEN? A big DUH!

I would like to propose that DUH stand for Darwin's Unbelievable Hoax!

FB: I myself believe in life itself and not exactly how it was originally created. Everyone is entitled to their own opinion.

Explain exactly what it means to "believe in life itself." Sounds impressive, but it is meaningless. So if someone asks you where life came from, do you simply respond, "I don't know. I just believe in life itself. End of story?"

Yes, everyone's entitled to their own opinion. Which means I'm entitled to tell you your opinion is wrong. And then demonstrate why. Then it's up to you to either change your opinion, or stubbornly adhere to it because you can't admit you're wrong. In the end, you will look the fool, not me.

FB: Evolution doesn't recognize color, so how can you say it is a racist concept? Camouflage is an example of how evolution uses colors to aid survival.

Scientists have already answered your first question; they recognize full well the racist implications of

evolution. Darwin's own *Origin of Species* has a
subtitle (you rarely hear the subtitle quoted) that
includes "the preservation of favoured races." Of
course, his white genteel race was one of the
"favoured" ones that needed to be preserved. His
attitude toward the Patagonians was that they were
sub-human savages.

The camouflage issue is yet another demonstration of
the stupidity of the concept of "survival of the fittest."
How could evolution "know" it had to produce
phenotypes that would be able to camouflage
themselves in their environment? There must have
been literally millions of phenotypes produced till
evolution "hit" on the right ones each time, and in the
meantime the living things that did not have
camouflage were not surviving, because they didn't
have camouflage, right? Can anyone explain how
things that were dying without camouflage led to
living things that do have camouflage? And it's not
like the camouflaged things can't die of other causes.
A man can go hunting in full regalia so that he can't be
seen, and he can die of a heart attack or be buried in
an avalanche. A lot of good the camouflage would do
him. Same with an animal. It could be sitting there
just reveling in how great its camouflage is and
suddenly a rock falls on it and kills it. Extreme
examples, but the point is camouflage is not 100
percent protection from anything.

FB: You have no proof that an afterlife exists.

You have no proof that an afterlife *doesn't* exist. Given
that someone did return from the dead (which I

assume you would not wish to believe despite plenty of eyewitness testimony), that's enough proof for me.

However, let's talk about proof for a minute. You have no proof that there is extraterrestrial life, but I'll bet you and a million scientists believe there is. You have no proof that evolution actually happened the way they say it did, but I'll bet you believe it. You have no proof that intelligence can evolve from inert matter, but I'll bet you believe it did. You have no proof that there are other universes, or that the Egyptians actually built the pyramids, but you choose to believe it (someone once proposed to me in all seriousness that apes actually built them - can you PROVE they did not??). There are lots of things for which there is little or no proof (evolution being one), but people still believe in them. People believe what they want to believe, and don't bore them with the facts.

FB: Why did the universe have to have a start? Don't you believe that by faith?

That's an easy one to answer. First, you would have to have GREAT faith to believe it did not have some kind of beginning, since "it exists."

Second, even the scientific evidence indicates that the universe had a beginning. If it did not, there would not be two laws of thermodynamics which state that the total energy in the universe is a constant, but the energy available to perform work is running out (entropy). Therefore the universe had a beginning and that can be demonstrated scientifically. We don't need to believe it by faith alone.

What we need to believe by faith is HOW that beginning came about. And it is not logical, given the scientific evidence, to believe that the complexity and structure in the universe, be it galaxies or atoms, came about by random chance processes. There is clear and obvious design present in all things from the largest to the smallest. Scientists recognize that, or there would not be scientific laws. Some of them just don't want to acknowledge that it could not have come about from nothing with no intelligent input, which is a demonstration of their own lack of intelligent discernment of the facts.

FB: Evolution explains everything. Without it nothing makes sense.

The truth is that evolution explains NOTHING. We humans invent the explanations and then fit them into the theory and say, "See! Evolution explains it!"

FB: No scientist would say the universe came from nothing. Where did you get that from?

You're wrong. Alan Guth for one, the inventor of "inflation" theory, realizing that his original "particle" must have come from somewhere, actually suggested that everything started out from... are you ready?... NOTHING. Ok, time for science to admit that every scientific law in the universe would be violated if that were true. Who flicked the switch? NOTHING did. There you have it. Other scientists have followed with the something from nothing theory and its adherents are a growing number.

FB: Isn't it interesting that many evolutionists still use B.C. and A.D. when talking about recent history?

Actually evolutionists and others have tried eliminating B.C. and A.D. and replaced it with BCE (Before the Common Era) and BP (Before the Present), neither of which, interestingly, is based on a fixed reference point. That goes well with evolution, where you can add or subtract millions of years at whim and nobody bats an evolved eyelash (you know, of course, that evolution "knew" we needed eyelashes to keep dust and dirt out of the eyes that evolution knew we needed to evolve to be able to see.) If they do still use B.C. and A.D., it's just because historically they've been used as a reference point, not because the users necessarily accept the existence of God or Christ.

FB: Where did the concept of evolution originate?

We can trace it as far back at least as far as the Greek civilization, who proposed a "ladder of life" or better, a "chain of being" that led from less to more complex beings, with man at the apex. Darwin was by no means the inventor of the idea that living things evolved; he just systematized the concept and gave some pitiful examples (which have been falsified at this point, though the textbooks haven't caught up with that truth) to supposedly demonstrate it.

For example, his finch beak research allegedly demonstrated evolution, but in reality it does no such thing, as his examples are now known to be nothing more than adaptations which arise from genetic information that is already present, and do nothing to explain where the information to make a beak (and put it in the right place on the finch's body where it would be most useful - a finch with a beak on its butt or wing wouldn't be able to use it very well, now would it?) came from in the first place. The finch beak is just one more example of evolutionists using change and adaptation to "prove" evolution, which they do not at all.

FB: What about mutations? Don't they demonstrate evolution and survival of the fittest in action? Elephants have evolved large ears to dissipate heat for example, and some mutations in sheep allow them to eat seaweed.

Mutations act on genetic information that is ALREADY present in the genome, and do nothing to explain the original source of that information, nor the transmission of it. Mutations are nearly always harmful, and NEVER increase information. They use what's already there in the genome. They duplicate it. They corrupt it. They don't add anything useful and new to it. That's a known and demonstrable scientific fact. Yes, fact. Mutations can't be used to explain evolution.

Your feedback is a perfect example of what I spoke about previously. Even though evolution really can't be used to attribute "purpose" to anything, you'll take an elephant's ear, see that it is designed in a particular way to perform a particular function, and then you'll affirm that evolution did it, as if evolution knew an elephant needed ears that could dissipate heat. (Not all species of elephant have large ears, by the way.)

Now show me in the fossil record where all the stages are in the evolution of the heat-dissipating elephant ear. And while you're at it, show me how evolution knew what size to make the ear, where to put the ear, how to make sure that blood would circulate in the ear to dissipate heat, how evolution "knew" that blood could even do that, how evolution managed to pass on the ability to make ears to progeny, how evolution knew what muscles the ear would need to allow it to flap, how evolution knew to put skin on the ear and cartilage in it to give it form and structure, how evolution knew that the skin had to allow heat transfer, and how evolution knew just how to hook it all up to the elephant's brain (which evolution knew just how to evolve) to make it all work. Oh, and by the way, evolution also knew that the elephant would know it was too hot, of course, otherwise the elephant might fry up and not be able to evolve, right?

The fact that sheep adapted to be able to eat seaweed means nothing if the machinery to then put that digestion to use is not there. In other words, did the ability to assimilate nutrients from seaweed also suddenly "evolve" out of nowhere? If you THINK about your argument, it has everything to do with adaptation using machinery and information that was ALREADY in place, and has NOTHING do to with the question, "WHERE DID SHEEP COME FROM IN THE FIRST PLACE?" Where did the sheep's digestive system and all that's connected with it come from in the first place?

Some of the best survivors in evolutionary history have been bacteria, correct? They have no wings. They're not big. They don't have sharp teeth. Imagine that, and they still survive to this day! Of course, the evolutionists have a contrived story to explain that one too. Funny how some bacteria just marched to a different drummer and decided that it was time to move "onward and upward" while some were content to just remain plain old bacteria.

How about your senses of smell and taste for another example? Are they the products of blind chance mutations? So evolution accidentally found out that you needed to drink liquids and be able to smell whether they were dangerous or not? While it was finding it out, you'd have been dead. And dead things don't evolve.

FB: Radiometric dating proves the Earth is billions of years old.

You're absolutely wrong. Nobody can "prove" any such thing. Radiometric dating is just ONE method of dating things, and it is based completely on ASSUMPTIONS. There are other ways of estimating the age of the Earth that do not agree at all with radiometric dating results, and even those results are questionable. Often if a "date" doesn't fit the presupposed age of a rock or other object, it is thrown out, and the object is re-dated till a result comes up that's closer to the expectations of the scientists dating the item. Some other methods of dating include extrapolating the human population backwards, calculating the amount of fluvial deposition of salts into the sea, how fast galaxies are winding up, etc. Those all give differing dates for the age of the Earth and universe that conflict with the older ages given by some rock dating.

FB: [Writer alleged that deer evolved the ability to run quickly as a survival tactic.]

That deer argument fails from the first sentence, because before deer could "run fast" how did they survive? And if they were surviving without running fast, then they didn't need to run fast to survive. That's a perfect example of the stories evolutionists conjure up to try to explain things to the unthinking masses.

FB: Christianity claims that each individual species was created discreetly, which is clearly not true.

"Christianity" does not claim any such thing, neither does the Bible nor creationists. You are grossly misinformed on that and need to do further research on it. It's not my desire to get into that here.

FB: What is the source of consciousness, and how can you deny that consciousness evolves over time, as man has become more and more aware of his surroundings and able to control them?

"Consciousness evolves" is a good example of the abuse of the term "evolution." If you mean "changes" then we're in agreement. If you mean that consciousness created itself and became self-aware on its own, and is developing something new all the time, then we part ways. Besides, "consciousness" is a very subjective thing, so you can easily say it "evolves" without being challenged. How can anyone prove otherwise? First you have to define "consciousness," though, which you've not done.

FB: Some starlight has taken billions of years to reach us, which proves that the universe cannot be young.

Concerning light taking "billions" of years to get here, that is another (cosmic) evolutionary fallacy. It is quite possible that we will discover some mechanism that refutes that notion (there's MY faith in action!). Riemannian space is one that has already been proposed, though I'm not going to pretend I understand its mechanism. Another is called "cdk" where "c" is the speed of light, and the proposal is that it was much faster in the past than now, and hence the speed has decayed (dk) over time.

This idea fits well with a "tired" redshifted light reaching us, as red light has less energy than violet light at the other end of the spectrum. As an amateur astronomer, I find it laughable that a light ray could travel unobstructed for BILLIONS of years and arrive on my retina with the same energy it had when it left its source. Also, if you notice Hubble and other deep space photos that supposedly show the initial stages of the universe, you'll note that there are already fully formed galaxies there, not galaxies in the process of formation, as we'd expect to see if the cosmic evolutionists were right.

FB: Creationism is all about a magical being creating things by magic powers. Creationists can't explain

where all the different species of cats came from, for instance. Evolution explains it perfectly.

Let's go over this. In order for evolution to have "created" various cats, first it had to start with nothing. The nothing then became something. Let's call the something molecules. So molecules magically appeared from basically nothing. Ok, then molecules got together and started forming stuff. Then they came to life. Magically. Then they figured out how to reproduce. Magically. Then they got bigger and started forming eyes, ears, whiskers, teeth, tongues, lungs, kidneys, nerves, digestive systems, organs that could say "meow," paws, and hair. And the neat part is, all those parts were magically organized and put in just the right places in a body covered by skin that was able to move, breathe, say "meow" and so on! Isn't evolution just...., well..., MAGICAL???

Ok, so now we know how evolution "made" all those cats, right? And THAT's SCIENCE. REAL science.

If you believe in creation, an original cat was created within which was contained all the genetic information to allow various species of cats to arise through genetic recombination. That's exactly what we see happening. With dogs, it was likely a wolf-like animal, and other dog descendents began to express some of the variety that was contained in the original genome.

However, with descent comes LOSS of information. Thus a cat or dog that is descended from the original would have LESS genetic information in its genome. And that's exactly what we see. We don't see NEW information arising, as would be necessary for evolution to take place. We see selection working on what was ALREADY there, and producing various expressions of genetic information content (phenotypes), without producing any novel information. Evolution only works in a downward direction (loss of information). Hence, it doesn't work at all, because for things to evolve from nothing to everything requires massive amounts of NEW information.

Oh, and by the way, if you happen to see any of those cats turning into something besides cats (which would certainly help the cause of the evolutionist!) please let us know!

FB: Can you give an example of how a mutation would corrupt information? We've actually seen examples of short-lived new information arising in flies and bacteria, and bacteria developing the ability to digest nylon. An understanding of how they adapt and evolve helps us find out how to combat diseases.

Ok, take the word "EYES." You know what an EYE is, right? Ok, let's take that information and alter it to "SYEE." You don't know what SYEE is? I don't either. How about "YSEE?" Or let's remove a letter, and just come up with "YES". Well, VOILA! We've created NEW information, right? Wrong. We've just altered what was there. Ah, but yes, we HAVE created a new word! However, in the context of the old word, the new one is USELESS. If you want to make EYES to be able to see with, YES isn't going to help much.

Let's take another example: the word GIFT. In English, we know what a gift is. You know what GIFT means in German? Poison. So, there's more than just altering existing information. We have to have a source of information. We need a code. We need a mode of transmission. We need a mode of interpretation. And finally we need a mode to be able to apply the information to perform something useful. What I'm saying is if you want to give me a "gift" you'd better be talking about the English version.

So, concerning any short-lived "new" information you're talking about in flies and bacteria - was it useful information? Can you give us more detail on how that supports the evolution religion? If it was indeed "short-lived" it must not have been very beneficial. I know of some wonderful "proof" of evolution in a fruit fly that developed an extra set of wings by duplicated genetic information. The wing information was already there and just duplicated. The fly was enabled to develop a second set of wings. The fly died. The extra wings were useless. Nothing new was created. Just a second set of wings using a blueprint that already existed. And it was useless.

The nylon digesting bacteria (which I've already discussed in this section) have been explained away long ago. I could cut and paste a huge refutation of how that example supposedly demonstrates evolution, but you'll have to do your own research. Try going to Creation.com and type "nylon" in the search engine. Do a little research.

I love stuff like that. Ha ha stupid creationists! We found some bacteria that can digest nylon, so that explains where human legs that wear nylon stockings came from! In other words, because we found a bacterial strain that can digest nylon, WHOOPEE! we now have PROOF POSITIVE that bacteria turned into Barbara over a few million years! Not that evolution requires a leap of faith or anything like that. This is just another example of how badly evolutionists have to strain to come up with examples that supposedly bolster their position. If we can just show that bacteria "evolved" the ability to digest nylon, that will prove that bacteria also turned into people.

The fact is, the bacteria already contained the machinery to develop the ability to digest the components of nylon. Bacteria have demonstrated an amazing spectrum of abilities to withstand and survive extreme environments. But they're still bacteria. Do you get that?

Bacteria don't evolve. Epidemiologists do not use evolution to determine what bacteria are going to do. They don't have any way of predicting how a bacterium or virus is going to "evolve" or we'd have plenty of vaccines on hand for future evolution events. What they DO work with is various already extant strains that for some reason or other become dominant for a period of time. Nothing new has been created (even in the case of mutations). The strains are already there, just waiting for the opportune time to fill a niche. Again, nothing to do with evolution. Medicine would, and will, proceed forward just fine without it. In fact, because of evolutionary beliefs in things like vestigial organs, in the past people have undergone unnecessary dissections to remove organs that were either misunderstood, temporarily unused, or just lying dormant. Every so-called vestigial organ (I've read at one time there were as many as 180) has magically found some purpose over the years.

FB: Is the Bible the literal truth word for word? Why believe in a book written by men thousands of years ago? They didn't know what we know today.

Is the *Origin of Species* literal? I mean, maybe Darwin meant it allegorically. Why not lets attack it critically like they do with other religious books? Or is it sacrosanct and above criticism?

No, you don't have to believe a book written by men thousands of years ago that has corroborating evidence to support it in every discipline known to man, from science to archaeology, and more manuscript evidence to prove the fidelity of its transmission than any ancient text. It's much better to believe a debunked book that was written by just one man 150-some years ago, or to believe some 30-year-old scientist who takes some smashed bones (a-la the recent "Ardi" skeleton) and tells you all about what some "pre-human" was like 4 million years ago. I think it's much more logical to believe that than some book that has been around so long. After all, the longer a book is around, the less true it becomes, right? I mean, don't books become untrue with time? I always thought so. Therefore, if an old math book says 2 plus 2 equals 4, we shouldn't believe it because that's an old math book. Or if some old history book says the pyramids were built by the Egyptians, we shouldn't believe that because it's an OLD history book, right? Yes, I see the logic in not believing what a book says just because the book is old. Thanks for pointing that out.

What you demonstrate is that you, like so many others, choose to believe the most negative information that supposedly supports your own belief, instead of actually researching whether that information is true or not. Just as one example: since darwinism hit the world, more people have been killed by those who bought into that philosophy than in all the "religious" wars throughout history. Millions have died in the name of darwinism. There are lots of books out on that. Try Dinesh D'Souza's *What's So Great About Christianity?* for starters, then read some of the late books out on darwinism's influence on Hitler, Stalin and the like.

FB: Some day evolution will be proven beyond a shadow of doubt. Just because we don't have all the facts now doesn't mean we won't some day. Evolutionists don't make up stories to cover the holes like religious fanatics do. We fill in the holes with facts until they're all filled. We don't believe in science fiction like the Bible teaches.

You are demonstrating your "faith" in a false system by making the common excuse that sooner or later evolution will be proven. Well, it's now been over 150 years since Darwin wrote his Evolution Bible. Isn't that long enough for some "proof" to show up? What you're doing is like the loonies who don't believe we landed on the moon. If we just wait long enough, will we find out they're right? No. Because they're not right. So, no matter how long you wait (and of course, in evolution, TIME and the waiting game are very important, aren't they?), the fact is that evolution is never going to be proven because it simply didn't happen and cannot be tested and repeated. Of course, you might want to wait millions of years, but

unfortunately you won't be around then to know
evolution was finally proven, will you?

You didn't do a very good job of convincing me that
evolutionists don't make up stories to cover the
"holes." Sorry, but that's exactly what they do. They
will accuse creationists of using the "God of the gaps"
but they have their own god of the gaps, and it's called
Time. Just wait long enough and sooner or later the
evidence will show up, right? No, that's not going to
happen. It's time to abandon evolution as a viable
theory. It's not, and never will be. There's enough
DISproof now.

As far as my pointing out what's wrong with
evolution, what's wrong with that? If I'm wrong about
what's wrong, then show me why I'm wrong! What I
believe has nothing to do with it. What you believe
does. If you believe in evolution and the best you can
come up with is a bunch of holes you need to fill,
something's amiss there, wouldn't you agree?

Since you are obviously a Bible expert too, I think
YOU should write the book about all the holes in the
Bible. Of course, you will make excuses for the holes
in YOUR religion, but if there are "holes" in the Bible,
imaginary or not, we can't make any excuses for those,
now can we? I mean, suppose a Bible believer says
there are "holes" in what he believes, but just because
they're not completely resolved, that doesn't mean it's
wrong. Would you agree with that statement, or does
it just apply to your religion of evolution?

"Science fiction" is the best term you could use. That
is EXACTLY what evolution is. In fact, it can be
demonstrated that the literary genre of science fiction

really did not become popular until about the time
darwinism became popular. Then all of a sudden you
had extraterrestrial life forms, and so on, and science
fiction went wild. Alleged science "facts" followed,
with the "canals" on Mars and such, supposedly
demonstrating that there was life on other planets. Of
course, we now know that's false, but did the
fictionists give up? No way! Now we have SETI,
which sucks untold millons of dollars, hoping to find
some "sign" of intelligent life out somewhere in the
universe, while its adherents don't have enough sense
to recognize the signs of intelligent design in creation
right here at home.

**FB: It seems to me that evolution is just a long made
up story that will not get you anywhere in life.**

"A long made up story that will not get you anywhere
in life." Great way of putting it! That is so true, and is
another anomaly concerning the followers of the
religion of evolution. WHAT HOPE DO THEY HAVE?
None whatsoever! You're born. You live a few years.
You die, and become fodder for future evolution.
What a WASTE of life. No purpose. No reason. No
hope. No future. All the questions they have will
never be answered. All the dreams they dreamt will
go to the grave with them. You think they stop for a
minute to think about any of this? I doubt it. Many
can't see past their noses. It's the "now" that matters,
and that's all. You can do what you want, and there
will be no consequences. Too bad they're wrong and
will find that out the hard way. Foolish people.

FB: Evolution is based on facts and reason, not faith. Science is the only real source of truth, not religion.

Say what? You believe in "reason" and don't believe in evolution by "faith" because you watched it all happen before your very eyes, did you? Well, feel free to enlighten the rest of us. Science is the only source of "truth"? That's about as foolish a statement as a blind person like yourself can make. If scientific "truth" is always being modified and falsified, then explain just how it is "truth" and define "truth" while you're at it. If all our brains are is a bunch of modified chemicals, how do you even know that what you think is "truth" is true? Maybe my "truth" is different from your "truth," but we're both right. Or maybe your "truth" is WRONG and mine is right. Try THINKING about this stuff and you'll be amazed how your views will change.

Ok, let's be honest, so I'm not accused of dishonestly discrediting your statement. Honestly, now, define "truth" for me. Honestly, now, explain how if "science" is the only source of "truth" and scientific information is continually changing and being discredited, explain to me how I'll recognize truth when I see it. If I tell my wife "I love you" but can't prove that scientifically, does that mean I've not told the truth? No? Then there must be other "sources" of truth in the world, otherwise we have to accept your statement that there are not as an absolute truth in itself, but since you can't prove scientifically that your statement is true, it may well be false. Got that? If not, let me put it another way: You're saying that "the only source of truth in the world today" is science, but that statement itself cannot be proven scientifically.

You've backed yourself into a philosophical corner there.

Can you prove to me that the results with another person's brain are actually the same repeatable, quantifiable results you're having with your own? You have not explained how an evolved bunch of chemicals determines what "truth" is. Suppose my evolved chemical brain sees green when yours sees blue - color blindness. Which evolved group of chemicals, if any, is seeing the correct color? And based on what criterion? The notion that "science seeks truth" is without foundation unless you define what "truth" is.

FB: So you believe a book [the Bible] that teaches that incest is ok? Science teaches that incestual relations corrupt the human genome. The Earth would be populated much faster than creationists say if we started with two people.

Oh, here we go with the "incest" argument. I suppose that all your ape ancestors had a code of morality where they wouldn't dare touch one of their ape siblings or kids, eh? Incest is only incest because God declared it to be so, not because of anything man invented, and He did so for the very reason you mention. As we got farther from the original sinless creation, we became more corrupt, including in our genetic makeup. If it were up to us, we'd be having sex with every living thing, including apes.

Did Darwin write a book about morality? I guess not. If all we're here for is to pass on our genes and have a good time, who are you or anybody else to say I can't

have sex with whomever or whatever I please? If you
say it's "wrong," on what basis? Just because YOU
think so?

And you are also absolutely wrong about how long it
would take to populate the earth beginning with two
people. There's plenty written on that. Do the
research.

**FB: Can you give some evidence for a young Earth
and universe?**

There is a lot of evidence for a young universe, and
plenty can be found online if you just look up "Young
Earth Evidence," rather than my parroting it here.
One of the best arguments for a young Earth is the
population question. If evolution were true, the Earth
should be covered by pre-humans and humans by
now. Rather, if you extrapolate backward from the
current population, taking into account various causes
of death such as famines, wars, pestilences, etc., you
come up with a population that's only a few thousand
years old. Further, written history is only a few
thousand years old. Before that it's all conjecture.

**FB: What are your thoughts on the flaws in carbon
dating techniques?**

Not only carbon dating, but all dating systems are flawed.
Carbon dating, for one, is only supposed to be accurate to
around 50,000 years. Oh, on a side note, carbon-14 has
actually been found in fossils that are supposedly "millions of
years old" but it's always attributed to contamination or some
other factor - anything BUT the fact that the fossil is NOT
millions of years old.

All dating is based on the following assumptions:

1. The original amount of radioactive parent product is known. However, it can't possibly be known. It's assumed.
2. There was no daughter product (like lead from uranium decay) present in the sample originally. This too is an assumption and a wrong one.
3. The rate of decay has always been constant. As every evolutionist knows, nothing in this universe is totally constant, except for when the evolutionist needs it to be so. Various factors are known to affect radioactive decay rates.
4. Often, the "age" of a rock is assumed before the rock is even dated. If the rock came from a particular formation or stratum, it is assumed to be a certain number of millions or billions of years old to begin with. So the dating folks conveniently end up with dates that more or less "match" what was expected. It's also well known that fossils are dated by the rock milieu in which they're found, and rocks are often dated by the "index" fossils found in them. Sound like circular reasoning? That's because it is.

In any event, all estimates of age are just that - estimates. Further proof of that is the margin of error that dates are usually given. Suppose I went to my banker and asked how much I had in my account, and he said $1 million plus or minus $100,000? Only evolutionary paleontologists can get away with such numerical legerdemain.

FB: I heard about some *T rex* bones that were found that still had pliable tissue present in them. The scientists were astonished that such tissue could be pliable after tens of millions of years! You'd think these educated geniuses might have concluded that maybe the bones weren't as old

as they said they were. It seems that the more educated people become, the more blinded they are.

Yes, I'm familiar with the *T. rex* bones. Blood cells were also found in them. I, as others, have commented, like you did, that these folks are so blinded by their *a priori* commitment to evolutionism and millions of years that they won't even consider the possibility that the bones are YOUNGER than their mythology requires them to be. Their reasoning is this:

Wow! How could blood cells and soft tissue have lasted MILLIONS OF YEARS!?

Instead it should be: Wow! Maybe these bones really AREN'T that old!

But since that might mean having to question their cherished fanatical evolutionary religious belief, they won't go that route. You'll notice that the hullabaloo around that discovery (which I believe was originally published in *Discovery* magazine) quickly died out.

FB: If you are a Bible scholar (even though you've kept this pretty secular so far so I'm not sure if you are or not), can you explain how the stars can be so old if the Earth is only 6000 years old?

I don't think by now it's any secret where I stand. However, that's still irrelevant as far as the evolution question goes. The plain fact is, evolution does not happen, never did, and never will. The "age" of the stars is an age old question (dumb pun intended), and I've touched on it already, but let's hit on some important points again:

First, a light year is a measure of DISTANCE, not time. This is important, because it's the distance light travels in one year. So the variable here is the speed of light, which supposedly doesn't change. But that has been seriously called into question as experiments of late have shown the speed of light can vary. Why everything else in the universe should change (and I don't mean evolve - the evidence and scientific laws demonstrate that the universe is in a state of decay) except the speed of light makes no sense. An Australian named Barry Setterfield has attempted to demonstrate a principle called cdk (c being the speed of light, and dk meaning decay), namely that the speed of light has decreased with time. So, at the beginning, perhaps the speed of light was near infinite, so that light was able to traverse enormous distances almost in an instant. In fact, even evolutionists have come up with the same idea. Alan Guth proposed the inflationary theory, in which the universe, in the beginning, expanded way faster than the speed of light.

Second, and this always gets my blood running hot - there is a philosophical DISCONNECT between distant starlight and evolution. Here's the problem. You're saying the following: Starlight takes billions of years to reach us, therefore life evolved over billions of years. See the *non sequitur* there? There is no connection whatsoever. If something can't happen in one year, it's not going to happen no matter how much time you add to it. A nickel won't change into a paper dollar in a year or a billion years. So if life did not, and cannot, evolve, it's not going to do so no matter how much time you add. But the evolutionists NEED time, because it's their smokescreen behind which to hide. They KNOW evolution can't be true, but sure enough if you say it needed millions of years, nobody can test or prove it, so your case is closed.

Third, there are other possibilities regarding distant starlight in a young universe. Have you ever looked at a Hubble telescope deep space image, where they say there are galaxies almost from the beginning of the universe? Did you notice those galaxies are FULLY FORMED? They're not "becoming" galaxies. They ARE galaxies. We should expect rather to see blobs of gas turning into galaxies if indeed we're looking at the beginning of time and matter. But we don't. Nor do we see star formation. That's another evolutionary fantasy. We hear about areas of "star formation," but has anyone actually seen a new star come into existence? No. We do see plenty of stellar decay, though. So I believe the light we're seeing is either not as distant as we think (it's not possible to measure it directly - the distances are based on assumptions, too), or there is some warp to space or space/time that allows distant light to reach us more quickly than it is currently assumed.

Again. the measure of billions of light years is based on *assumptions* - mainly the brightness of "standard candle" stars like Cepheid variables. Scientists don't seem to have learned the lessons of history that turning assumptions into dogma gets them into trouble. Every age of "scientists" or "naturalists" or "philosophers" or whatever has thought that their "wise ones" had the right answers, only to turn around and find out they were completely wrong. Even evolutionists must postulate an original expansion of the universe that was much greater than the speed of light. So that would give something the appearance of being billions of light years away, when the light actually traveled that distance much more quickly in the beginning so the light is not that old. Or there may be some other mechanism by which light traverses the universe more rapidly than we think. I repeat, the galaxies we see at the so-called edge of the universe are GALAXIES. Fully-formed.

It's like the evolutionary book I just read about the fabulous fossil finds at Riversleigh in Australia. The writers go on and on about all the various species that were found, and how they "might" have evolved, but the FACT is they are all FULLY FORMED and FULLY FUNCTIONAL animals. They're not part-animals, or deformed animal monsters that are turning into kangaroos, platypuses, etc. They are fully recognizable - many appear to be the same as currently living species, showing to me and those like me that they are NOT millions of years old. The point is, they are FULLY FORMED and functional, not evolving into or from anything. Find the ancestor of the platypus!

Regardless, no matter how much time you add, evolution didn't happen, doesn't happen, and will not be happening any time in the future.

FB: How is it that biology and geology agree on the age of the Earth, but you say they're wrong?

Geologists and biologists don't agree on anything. Just like all branches of science, you have agreement between some and disagreement between others. So there are numerous scientists who believe in billions of years and look for "evidence" to support it, and there are numerous scientists who DON'T buy the story line and look for "evidence" to support it. Funny thing is, we all have the same evidence. It's all in the interpretation.

Think about your question. Do you realize how many times the supposed age of the earth has changed over the past 150 years or so? It's gone from thousands to millions to billions. Why? Because Time is the evolutionist's only ally. Their reasoning is that any miracle (they won't use the word, but that's exactly what it is) can happen if you just have enough Time.

FB: Scientists do not fudge their results.

If you really don't think scientists fudge their results, you're either deliberately blind, or naive. Scientists are human like the rest of us - and fallible - especially when money, or reputation, etc., are involved. In fact stories are always coming out about how scientists do just that, and evolutionary history is filled with them - Piltdown man, Nebraska man, the peppered moth - do some research. Or read *Icons of Evolution*, by Jonathan Wells.

This brings to mind the issue of peer review too. Scientists definitely allow their biases to rule when it comes to that. Do a little research on bias in peer review. That's why I allege that peer review in general is blarney too.

FB: Just how old *do* you think the Earth is?

I do not think the Earth is more than about 6,000 years old, which is about when man shows up and written history begins. If you want to believe man was sitting around twiddling his fingers for a few million years and then around 6-10,000 years ago he suddenly evolved the ability to cultivate plants, build cities, write, communicate, and on and on, feel free to believe it.

I recently saw a huge book entitled *Chronicle of the World*, which is (obviously) a history of our world. It has 1,294 pages total. Of those...

- One page chronicles the "evolution" of the world.

- One page chronicles the "evolution" of man.

The rest is, well, *actual* HISTORY. Take it from there.

FB: I almost laughed myself [silly] reading your [book]!

Well, I'm glad to see someone got the humor as I intended it! I do think evolution is laughable, and actually might do more writing on that line. I think we take it all too seriously. Did you see the recent TOOTH that was found in Israel that's going to (once again.... yawn, yawn) change the story of evolution? I loved the fact that the scientist who found it was named Gopher. Anyhow, I really do think this stuff is comical, and even more so that people believe it! Comical, yet sad.

FB: You are [not very nice word removed] retarded. I'm so glad you're kind is dying out...evolution and whatnot.

Thanks for the intellectual input. Never heard of [your university], but I guess you're getting a good education there.

[Author's note: I threw this one in just as one example of numerous such emails I've received over the years (usually anonymous as cowards tend to do) from 'intellectuals' who have a viewpoint that differs from my own. I have also received many that are a bit more encouraging, such as the following.]

FB: Liked your book. I used to take evolution as undisputed fact until I read it, and it made me think.

FB: I'm not a scientist though I'm well educated and I agree with your position, but it seems to me that while macro

evolution is obviously rubbish, micro evolution does in fact take place.

Re: microevolution, keep in mind that it's not creating anything new. It's just minor changes that are better called adaptations, and that are the result of information that's already contained in the genes of the organism. The evolutionists will corner you and say that numerous "micro" evolution events add up to macro evolution. They do no such thing. For "macro" evolution to occur would require massive amounts of NEW information that was not already contained in the genome. Doesn't happen. Never did. Never will. They know we have that argument so are looking desperately for any example of "new" information arising from nowhere. The one or two attempts they've come up with have been pathetic, for instance the ability of some bacteria to digest nylon (which I've discussed elsewhere here).

FB: I would have thought that Darwinists would have given up by now. What is it that keeps this stubborn theory going, anyhow?

There are a number of reasons people hang on to Darwinism. Atheist Richard Dawkins thanked Darwin for making atheism "respectable." What does that tell you? The first, and main reason, is that people who think they can avoid a God to whom they'll be answerable turn to evolution to somehow "prove" God does not exist.

The second reason is that darwinism is considered to be "intellectual." If you want to be looked up to as being among the intellectual elite, you have to subscribe to darwinism, despite the fact that there are thousands of highly educated and, by the world's standards, successful people who recognize darwinism for the drivel that it is.

The third reason is that many people who have been backed into a darwinian corner are simply AFRAID to reneg on the theory because their status, or jobs, or the admiration they receive from others, are threatened. They're unwilling to stand up and say it's simply not true. It's cowardly in some cases, but I understand it in others, especially where one's livelihood is at stake. I've known teachers and scientists who did not believe evolution at all, but just could not come out and deny it for fear of ostracism or losing their employment. If you don't think that happens, watch the documentary "Expelled: No Intelligence Allowed."

FB: Can you explain why any reasonable scientist would consider Evolution to be good science? The very idea that environmental changes can be passed on genetically is absolutely crazy. The book, *On the Origin of Species by Means of Natural Selection, or the Preservation of Favoured Races in the Struggle for Life* classified the races of humans with whites and asians at the top and was used by the Nazis to justify the Holocaust. Liberals use the theory to try to eliminate God and put forth their agenda, never realizing that godless socialist society did not work in the Soviet Union and cannot work because of the lack of stimulus. Why is it that people ignore the sociopolitical implications of evolution?

No "reasonable" scientist would consider evolution to be true. Unfortunately "reason" doesn't really come into play here, though the atheists and evolutionists will often appeal to the word, as if they're the ones who are "reasoning" and the others are fools.

Scientists are people. People have biases, fears, alibis, motives, etc. Many scientists are simply afraid to admit evolution is false, and with good reason. Despite the fact that scientists will say they're "open minded" and that science is all about testing, repeating and falsifying, the fact is that if they have a bias, they're going to support it no matter what. For a scientist to have ascribed to, and taught, and defended, evolutionary theory, then to suddenly say it's wrong, just is not going to happen in most cases. There's pride. There's fear of peers. There's possible loss of status or employment. I and others have pointed out over and over that Evolution has been used to justify the most horrific crimes ever committed against humanity. Now, some will counter that religion has done the same. Two wrongs don't make a right, so that is no justification, and is also a gross exaggeration. Read Dinesh D'Souza's *What's So Great About Christianity?* for more on that topic. He crushes the opposition who argue that "religion" has been the cause of all the ills of history.

I have argued too that evolutionists really are hypocrites when it comes to being "green" and "saving the Earth." Because if what they believe is true, all the wonderful living things we see around us were the result of some fortuitous events involving poison gases, lightning, and extinction events. So why worry? We should be happy and excited about all the new life forms that are going to arise out of our polluted earth, right?

FB: I can't believe how fallable [sic] your [web]site is! It's ignorant people such as yourself that will help keep our nation from prospering in science and physics. Teachings such as this will help turn the U.S. into third-world nation. Get with the rest of the First Word [sic], dummy: EVOLUTION IS HAPPENING...DUH!!!!!!!!!!!!!!!!!!!!

Thanks for telling me how ignorant I am. Just a few suggestions, though. You spelled "fallible" wrong for starters.

You also have no idea what you're talking about, as evolution makes no contribution whatsoever to real science (of course, if you can demonstrate how we'd have never gotten to the moon or invented the auto without it, I'm game). The USA became a first world nation without any help from evolution, and no scientific discovery has needed evolution as its basis (ok, maybe the lightbulb somehow can be connected with evolution - let me think about it). Since the teaching of evolutionary science fiction entered our science curricula, science literacy in the USA has gone consistently downhill.

And your grammar and punctuation also needs work. Evolution may be happening, but it does not seem to be helping you much. If so, do let me know how.

FB: I must ask you your credentials... It seems as if you do not know much at all about the nature of the physical sciences. Combined with what I see as an abysmal knowledge of the experimentation leading up to the conclusions, as well as the scientific method as a whole, I am forced to say you are quite foolish.

I possess [here the writer impresses me with his educational qualifications]. Though I am by no means an expert, it is even apparent to an apprentice that you are quite foolish. You can't hide your lack of knowledge with sarcasm.

You need my "credentials"? I am a rocket scientist, physicist, medical doctor, lawyer, blue collar worker, police officer, pastor, nurse, sanitation worker, waiter, and on and on. My credentials have absolutely nothing to do with anything. I'm a man who has used the brain God gave me to sift through the garbage I've been fed and not accept it just because the "experts" tell me it's so. I'm a Galileo, or a Newton, or a Luther, or anyone else who refuses to submit to the party line, and rather do my own thinking and draw my own conclusions, even if they go against the tide of lies with which I'm being inundated.

I'm not impressed at all by your fancy "degrees," sir. What your degrees should REALLY do is humble you to the point that you realize how LITTLE you know about all there is to know in the universe. Your puny degrees might impress you and some of your peers, but just because you know a little extra about one or two subjects doesn't make you a god, nor an expert in everything. I've known people like you, and what you need is a dose of humility. It's how much you DON'T know that should impress you.

Now, since you were actually there, and witnessed the evolution of the universe, and can test and repeat it, be sure to do so for all to see, so we can REALLY be impressed by your credentials.

FB: You say Evolution is not necessary for any branch of science to advance. Scientists say otherwise. Why do you think you know more about their work than they do?

"Scientists" say lots of things. Other "scientists" disagree. They say the Earth is warming up, despite the fact that it's actually cooled the last few years. They once said the universe revolved around the Earth. Some of them say not to eat eggs. Others say eggs are good for you. Too bad you've made them your final authority. Not a good thing. Try thinking for yourself, please.

FB: You say that when an antibiotic is introduced, bacteria that are ALREADY PRESENT and have resistance flourish, where the normal flora is killed off. Right! That's how

**evolution works. Survival of the fittest - in this case, the
most drug resistant.**

You gotta love it! No matter WHAT I said, that would be how
evolution works, because the brainwashed see everything as
working *for* evolution. I pointed out that antibiotic resistance
has NOTHING WHATSOEVER to do with how anything
evolved or is evolving and VOILA'! to you, that's just how
evolution works! Just gotta love it. "Survival of the fittest" has
been amply demonstrated to be a tautology. You have to
define "fittest" and of course, those are the ones that "survive."
In other words, it's just another evolutionary word game that
explains nothing whatsoever.

**FB: Irreducible complexity says that a system can perform a
given function ONLY with all parts present, but each part
can perform a different function independently of the
others, or the system can perform a different function when
parts are removed, so the system is not irreducibly complex.
Irreducible complexity suggests that if a single part is
removed from the system, the system will be completely
useless for ANY function.**

Irreducible compexity is the argument that *to perform a
particular task,* all the parts have to be there and fully
functional. In other words, to use one of Michael Behe's
examples from *Darwin's Black Box,* to have blood clotting you
have to have the entire clotting system ready and functional or
else the whole thing fails. Without blood clotting, we would
not exist. So to take away a part of the clotting cascade
mechanism and say that that part could have originally been
utilized in some other capacity fails as an argument, because
what you NEED it to do (clot blood) it no longer does. Like
taking a carburetor and making it into a flower pot and saying

you've improved your automobile. Take away one part of the clotting mechanism, and we die. Dead things don't evolve.

FB: Imagine you were watching a slide show of pictures taken of a person everyday for their entire life. The change from day to day would be so gradual that you couldn't notice it between any two photos, but would notice it to a greater extent the further the two photos were removed from each other. This is analogous (in incredibly simple terms) to a complete fossil record for a species.

No it is not at all. The first flaw in your argument is that the person would show deterioration with time, not improvement. The second flaw is that it's observable and quantifiable. There would be no guesswork involved, no contrived stories. The FACT is, the fossil record shows nothing but fully formed, fully functional organisms. It's a well-known embarrassment to the evolutionary community that evolutionary "trees" show fully formed, fully functional organisms at the tips of their branches, but nothing in between. If your "film" scenario could be used to show evolution it should show millions upon millions of flawed, useless things evolving while evolution was "weeding out" the "stuff" that would not work and honing each living entity into the one with the most "survivability." In other words, there would be teeth showing up on knees, and wings on noses, and fingernails on tails, and lungs on ears, and so on, till evolution got it all right, and got the wiring to the brain all correct so it would all function together. Instead, we see a fossil record where functionality and form rule, not chaos.

FB: You can look at a rock and see that it HAS eroded. You can't look at a rock and see that it IS eroding. You can look

at a species and see that it HAS evolved, but you can't actually see evolution happening.

Well, now, that is very convenient. We're sure evolution is happening. We just can't see it. And we're sure it HAPPENED. We just didn't SEE it. Only evolutionists could accept such a thing. If a creationist came up with it, he'd be laughed into oblivion. In fact, if a creationist came up with a lot of what evolutionists proffer, the same would happen. Suppose a creationist said the Bible was written somewhere between 10 and 20,000 years ago? That's a margin of error that pales in comparison to what evolutionists come up with, but would never be acceptable if a creationist said it. I once did an Internet search for the age of the universe, and came up with claims between 8 and 20 BILLION years. Just stop and think about that a minute.

FB: Why is evolution not possible without abiogenesis?

The very reason for the existence of abiogenesis, which is a fancy word for the falsified theory of spontaneous generation, is that scientists have to come up with some way that life arose from inorganic material. In other words, you can't just start with a single-cell organism and take it from there. Inquiring minds want to know how a single-cell organism arose from inorganic matter. So you push the evolutionary timeline back further and further till you arrive basically at the fact that nothing was there in the beginning, and nothing somehow managed to make itself into everything.

FB: Cosmology has nothing to do with evolution. You keep confusing the two, which demonstrates that you really don't know anything about evolution.

Why are you people so dense? If you have done any reading
on the subject of evolution whatsoever, you would know your
statement is wrong. What you're trying to do is get around
the fact that you can't demonstrate how living things arose
from nothing, so you evade the subject by saying that's
cosmology, not evolution. Well, I'm not letting you get away
with it. If you can't show me where the material came from
that turned itself into living things, you're nothing but a
deceiver and blinded religious fanatic. I'm not going to let
you get away with making the starting point of evolution the
"first living cell" as if you then don't have to demonstrate how
that evolved to begin with.

**FB: Evolution does [sic?] claim that people make themselves
out of nothing (apparently you are unclear on the very topics
you claim to argue). Only Creation claims that things are
created out of nothing.**

"Evolution" doesn't "claim" anything. People do. What you
need to do is go back to Evolution 101. If people did not come
from nothing, then where did they come from at the very
beginning, before there was anything that could become the
something that became people?   This is a CLASSIC example
of the religious fanatical nature of evolution. You believe in
something blindly, and make every excuse for that belief,
despite all the contrary evidence and logic.

**FB: You do realise that not every organism that dies
becomes a fossil, right? It's quite a rare process. The fossil
record is chock full of so called "failed evolutionary
experiments." They're what we call extinct.**

You are not thinking. Things are going extinct in our own day
too. I think this is kind of funny, actually, how hypocritical

evolutionists are. They're crying and whining about things going extinct on the one hand, while on the other they'd say that if things DIDN'T go extinct, "new" living things could not take their place. Well, where are all the "new" living things that should be popping into existence to supplant the ones going extinct? And why, for that matter, get upset about pollution and such? Won't new living things just adapt and evolve to live with it? Didn't all living things start in a mixture of noxious gases at the beginning of time? So why worry?!

The fossils do not show "failed" evolutionary experiments. They show completely formed, functional living things, with eyes, bones, etc. They show things that died off for one reason or other, and were NOT replaced. A "failed" evolutionary experiment (and the record should show COUNTLESS ones) would be something like an eye on the butt of a horse, or an elbow on the forehead of a human, or a toe sticking out of a knee. THOSE would show that evolution was "at work" to figure out where would be the best place for those things to be. How about a stomach attached to a hand, or a mouth with teeth on the outside, running up the nose? That would be a *real* evolutionary experiment in the making. How is it that evolution "knew" just the right places to put everything? How did it know to put teeth in your mouth, and not on your feet? Have you ever THOUGHT about this stuff?

FB: Scientists all over the world agree that certain scientific principles are facts. Scientists in India agree that acceleration due to gravity is exactly what North American scientists say it is. Science agrees that Evolution is a fact. If you want that to change, then you better start coming up with some repeatable evidence. NO evidence support creationism or Intelligent Design - as demonstrated by your inability to actually provide any evidence.

A BIG LOL! to that one! Gravity is testable and observable, though no one knows what it is. Acceleration is observable. Evolution is not. Now, if you can come up with a repetition of the Big Bang, so I can see it turn into everything we know, and along the way you can show where life came from, where the first cell came from, how it began to reproduce, and how it turned itself into you and every other living thing, we'll get somewhere. If scientists everywhere agreed evolution was a fact, there would not be an ever-growing, ever-vocal group of 'em that say it's not. Try looking this up for starters; I'll bet you didn't even know it existed. It's "A dissent from Darwin," signed by hundreds of scientists who no longer believe darwinism's tenets hold water. **www.dissentfromdarwin.org**.

I'll pose the same question to you that I've posed to probably hundreds of other evolutionists. It's like the Flood question, which is, "If there were a global flood, what evidence would you expect to see?" I have NEVER gotten a straight answer from an evolutionist on that one. Evasion, changing the subject, blowing it off as a "dumb" question, etc., but NEVER a direct response.

So here's my question on design: "What evidence would you expect to see?" Thanks.

Now I'll tell you what evidence I'd expect to see if evolution were true. I'd expect TRILLIONS upon TRILLIONS of FAILED fossils in the record. I mean all sorts of evolutionary monsters, not fully-formed, functional entities. I'd expect to be seeing LOTS of evolution in action RIGHT NOW, that is, species changing from one to the other. I'd expect to see new life forms appearing all the time. I'd expect to see evidence that nothing can turn itself into everything with no outside help. I'd expect to see evidence that life could come about all on its own. I'd expect evidence that reproduction was a simple process that could just become more complex over time. I'd expect there to be "simple" organisms, and there are not. The more we look into the cell alone, the more complex it appears. I'd expect to see nebulous galaxies at the edge of the universe, and that things would appear to be becoming more organized, rather than falling apart all around me and throughout the universe (I'm talking about the NET entropy of the universe, which is INCREASING, so don't try pulling the "crystals" bit on me, because that one's been debunked). I'd expect to see billions of fossils continuing to form all over the world. I don't.

Want more? If Intelligent Design were true, I would expect to see a sudden explosion of fully-formed and functional living things in the fossil record (ever heard of the Cambrian explosion?). I would expect to see a lot of diversity in living things due to the creative process involved in making the original plants and animals. I'd expect to see an obvious purpose and plan in the overall ecological systems of the Earth, and purpose and plan in individual organisms, such as that the tongue is for tasting, teeth for chewing, stomach for digesting, intestines for assimilating, anus for eliminating. I would also expect to see a brain that controlled all that, along with all the other purposeful systems it controls. I would expect to find sentient beings who could investigate it all and discover that there really was a creator behind it all, but I'd also expect those beings to have the freedom to reject that notion, otherwise they would not truly be free, but rather automatons made to do their creators will regardless. Just for starters.

FB: Please provide evidence that Evolution has been falsified.

I've already done so. No matter how much evidence I provide, it's not going to make a difference to some people. It's out there. If you don't want to see it, you won't.

Mutations don't work; they don't create anything new, but tamper with and usually ruin information that's already there.

Natural selection only works with what's already there, and creates nothing new.

People don't create themselves out of nothing.

The fossil record gaps have not been filled, nor will they. The fossil record shows fully-formed, functional living things, not things "becoming" things.

Information, such as that found in genetic material, cannot arise on its own, without a source, nor is it useful standing on its own, without a code, mode of transmission, mode of reception, and the machinery to make it useful.

The universe is winding down (entropy), not up.

I can go on and on, but already have done so, as have so many others before me. If you don't want to listen, you're not going to. If you don't want to see, you're not going to.

FB: Science has nothing to do with philosophy or souls or theology. Science is about how things work.

Absolutely WRONG. Anyone who's been in the scientific arena for long knows full well that not only philosophy (worldview), but bias and, perhaps above all, ECONOMICS, play a large part in scientific "investigation" too.

And it's only about "how" and not "why" or "when?" Are you kidding?

FB: Science proves the lies of religion!

So science has disproved the existence of God? Even Dawkins gives God a one percent chance of existence (see the "Expelled" DVD) just to hedge his bet. On the other hand, given that the very basis of science is that something can be FALSIFIED, it seems much of science is about proving itself wrong! Of course, though, the excuse is that that's what science is all about, not that the unthinkable would happen and - GASP! - a SCIENTIST would actually tell a LIE about something. (That is, of course, as long as money's not involved...)

[Addendum from John: This same writer recommended that I read *The Greatest Show on Earth : The Evidence for Evolution*. When I found out that the book was by Richard Dawkins, I was disappointed. I thought it was going to be by a more interesting, challenging author. One of Dawkins' previous books, *The God Delusion*,was aptly named, only he's the deluded one. The popularity of the book pretty much shows the average person's depth (shallow) of thought. Dawkins is not a good writer, even less a philosopher, and it is one of the few books I've ever put down before completion, because I found it completely boring, thoughtless, and unchallenging. At least evolutionists like Sagan and Gould presented thought-provoking analyses of scientific and historical information. Dawkins just has a chip on his shoulder. If anyone would like to see Dawkins make a fool of himself, simply view his interviews in the videos *From a Frog to a Prince*, or *Expelled: No Intelligence Allowed*. But the guy just doesn't give up.

Atheists are among the easiest to back into a corner. Simply ask an atheist the following question: "Do you know everything there is to know?" The honest atheist, if there is such a thing, would answer, "No." If they respond "Yes," you should consider them to be delusional and end the conversation there.

Assuming a "No" response, you can then ask them the following: "Ok, then, if you do not know everything there is to know, is it possible that something you DON'T know just might be God?" The honest atheist, if there is such a thing, would answer, "Yes." If they respond, "No," they are both delusional and a liar. This is exactly how Ben Stein corners Dawkins in *Expelled*. Dawkins claims he is "99 percent sure" that God doesn't exist (leaving himself a little wiggle room, just in case, though that 1 percent is a rather big gamble in his case). Interestingly, he's 100 percent sure that evolution occurred, even though 99.9 percent of mutations are harmful, and the rest create nothing new (which is the point on which Dawkins is cornered in the other video [*Frog...*]). Stein then asks Dawkins how he ended up with the 99 percent figure, and seeing Dawkins squirm and try to worm his way out of that one is priceless.]

FB: I rely on experts in certain fields to explain things. You go to the doctor when you are sick and the mechanic to get your car fixed. I don't have time to deal with issues that have been explained many times to weirdos like you by people who work in these fields for a living.

Explain to me how someone can be an "expert" in origins. Do you know someone who was there who actually saw the Big Burp create everything from nothing? And they can explain exactly how it was done, eh? Start thinking for yourself, and you'll soon discover the "experts" aren't quite as savvy as either you or they think they are. Evolution is an invented story, and everything they do is an attempt to fit "science" into their mythology. You can be an expert in ANY field without having anything whatsoever to do with evolution. It's not necessary for the advancement of any field of science.

FB: Why am I not surprised that you would need "Truth" defined?

Because the evolutionist has no solid ground on which to define "truth." That's why.

FB: Last week, scientists announced a new vaccine against HIV. They tell us that an understanding evolution is essential for understanding HIV.

"THEY TELL US"? Whoop de doo for "them"! Have you actually looked into it yourself? If you did you might be surprised to find out that evolution is a totally unnecessary corollary to the understanding of any disease, HIV included. The fact that viruses adapt and change has nothing to do with where viruses came from in the first place, nor how they're able to adapt and change, which are mechanisms that are already in place.

By the way, have you ever seen a T4 macrophage virus? Mind boggling complexity! It looks like a lunar lander. Tell me how that evolved and we'll take it from there.

FB: Why aren't shared pseudogenes in chimps and humans very compelling evidence for a common ancestor?

This is just another example of straining to come up with the silver bullet that's going to kill off the competition and leave evolution the undisputed winner. Pseudogenes or not, you still haven't demonstrated where they came from to begin with, how they became functional, where the information they contain came from, etc. And if they are NOT functional, how does THAT support evolution? They're useless genes. So what? However, there is current evidence that some pseudogenes are in fact functional, but regardless the fact we might "share" some with primates is in itself meaningless unless INTERPRETED by evolutionists to somehow pervertedly support their mythology.

Regardless of what genes we "share" with primates, the genetic information expresses itself in completely different ways. Did you know humans and bananas share about 50% similar genes? Meaningless. It's how the information contained in the genes is expressed that counts, not the similarities in makeup. I like to use the example: GODISNOWHERE. That might say GOD IS NOW HERE or GOD IS NOWHERE. One hundred percent similar letters (makeup). Completely opposite meaning (expression).

So the number of genes, pseudo- or not, that we share with chimps or bananas is meaningless. Genes just happen to be the carriers of information. How that information is interpreted and expressed is what's important, and evolution has no explanation for either the source of the information in the genes, nor for its origin, nor for the complex mode of transmission and its expression in various phenotypes. You can't explain any of it by random, mindless processes.

FB: Thanks for the laughs while pointing out the obvious lack of reasoning behind the theory that so many these days call "fact". Evolution is not only un-scientific, but probably the biggest barrier to any real scientific progress in the understanding of biology in modern times. Evolutionists are - simply put - stupid, and must realize this before they

have any hopes of enlightenment. Thanks again for telling it like it is.

Thanks for the feedback. I wouldn't necessarily say evolutionists are stupid. They just have nothing else to hold onto, and refuse to consider the alternative because they're afraid of it, and afraid of peer pressure. Fear is the motivating factor behind any rabid, blind faith belief, and the evolutionists know they can prey on that. Keep your peers in fear of ridicule or looking "less intelligent" or backward, or whatever, and you've got them right where you want them.

FB: There are some very good explanations about various moments in evolutionary history. The details of how every single change happened are, like all history, impossible to see for sure (though if that's a valid argument against it, then any account of past events is false, because the only way to verify it is to look at the evidence that remains after the event), but we can look at the fossil record, which shows very clear lines that match with carbon dating and geological layout as being millions of years old.

Hey, I like it! GREAT MOMENTS IN EVOLUTIONARY HISTORY! Like, when men and women split apart and became men and women. I wish I'd been there to see it! Or when the first fish walked out of water, put on his sunglasses, poured himself a margarita, and said, "Wow, I didn't know the beach was so hot!"

The only clear lines the geological record shows are clear lines. Which means the strata were laid down successively and contemporaneously, not over eons. In fact, that's undeniable. I just came across an article about the Grand Canyon. The article mentions, and I had a good laugh over this, the two competing theories evolutionists have about its formation.

One says it's old and formed about 70 million years ago. The other says it's young and formed about 5 million years ago. See!?? The PROOF is RIGHT THERE in the STRATA! Except they're having a little difficulty interpreting what the strata seem to be telling them, but aside from that, I'd say it's a pretty exact science, wouldn't you? Well, there are other interpretations of how the strata were laid down, but since you've been brainwashed, you're incapable of evaluating them.

Also, carbon dating isn't used to date things that are supposedly "millions of years" old.

FB: You're seriously claiming that the only reason we believe anyone lived in caves is because of a few jaw bones? Do you ignore the numerous cave paintings, remnants of food, fire, clothing, and other refuse in caves throughout europe, africa, and the middle east? Not to mention very complete skeletal remains of ancients humans in these areas.

People still live in caves. Proves nothing. As for the paintings, I often chuckle to myself that they were probably just doodles to waste some time, and archaeologists then turn around and attribute some intense, often spiritual meaning to them. It would be fun to know the real reason people drew some of those things on cave walls (kudos to Gary Larson and his "Far Side" cartoons for stimulating my thinking on that!). Oh, and there are also drawings and carvings of dinosaurs, triceratops included, around the world. Have you heard about them? Of course, you'd have some lame excuse for why they could not have possibly been dinosaurs, because, as we all know, dinosaurs lived way before humans.

FB: I think I'm right when I say that there seems to be a huge disagreement on how eukaryotes evolved from prokaryotes, as well as how prokaryotes evolved from prebiotic soup, as well as how sex evolved. These are really great questions that people have dedicated their lives to researching. Current hypotheses point toward endosymbiosis and there's pretty good reason to believe in it: we are observing an endosymbiosis happening right now. If interested, please read "A Secondary Symbiosis in Progress" by Noriko Okamoto and Isao Inouye.

Re: endosymbiosis, it is by no means an explanation of how prokaryotes evolved into eukaryotes. All you do with that explanation is open up a can of worms. How did the symbiotic relationship begin, and if each organism was surviving without that relationship, how did symbiosis confer any survival advantage? How did the symbiotic organisms come into existence to begin with? Why is evolution almost always somehow, magically, diverted onto a path of increased complexity and functionality? Is there some magic "ether" involved in evolution that forces it in that direction? Can we actually, experimentally, extrapolate BACKWARD from the organelles contained in a cell, to the original "symbiotic" organisms, and is there actually any hard evolutionary evidence (fossils, etc.) that demonstrates the factuality of it?

What you're really doing is what Fred Hoyle, Francis Crick, and others did with the question of life's origin on Earth. Since they realized it was way to complex to have just come about by random processes, they invented "directed panspermia" which pushed the origin of life to "somewhere out there" in space, where it then came and "seeded" the Earth. They therefore felt they could avoid the question of where THAT life came from to begin with. And likewise, you can't explain how nuclei and organelles evolved, so symbiosis comes to your rescue, or so you (and they) think.

FB: I just did a Google search on the topic "Journal Papers on Biological Evolution" and received over 5 million hits. Is that enough research that has been done on it so far? Even if 1/10 of the papers are actually on the topic of the Theory of Evolution, that's still 500,000 peer reviewed, tested and accepted papers by the people who are qualified to discuss this.

I just did a Google search on "eat flies wrong" (left out a key word meaning "doodoo" so as to not be vulgar), and 2,590,000 hits came up, so guess what we're having for supper?!

Then I did a search just for "astrology" and TWENTY SIX MILLION hits came up, so I guess we'd better all go check our horoscopes, because that PROVES that astrology is scientific!

And that, my friends, will conclude our enlightening section on feedback. Hope you've enjoyed it!

--The Author

For Further Reading

Bates, Gary

Alien Intrusion : UFOs and the Evolution Connection
Creation Book Publishers, 2010

Behe, Michael J.

Darwin's Black Box: The Chemical Challenge to Evolution
The Free Press, 1996

Denton, Michael.

Evolution: A Theory in Crisis
Adler & Adler, 1986

Davis, Percival; Kenyon, Dean H.

Of Pandas and People: The Central Question of Biological Origins
Haughton, 1999

Dawkins, Richard

The Blind Watchmaker: Why the Evidence of Evolution Reveals a Universe without Design
W. W. Norton & Co., 1986

Ekers; Cullers; Billingham; Sheffer (eds)

SETI 2020: A Roadmap for the Search for Extraterrestrial Intelligence
SETI Press, 2002.

Gonzalez, Guillermo; Richards, Jay W

The Privileged Planet: How Our Place in the Cosmos Is Designed for Discovery
Regnery, 2004

Hunter, Cornelius

Darwin's God: Evolution and the Problem of Evil
Brazos Press, 2001

Jastrow, Robert   *God and the Astronomers*
                         W. W. Norton & Co., 1978

Johnson, Philip E   *Darwin on Trial*
                         InterVarsity, 1991

Kerkut, G. A   *Implications of Evolution*
                         Pergamon Press, 1965

Keynes, Randal   *Darwin, His Daughter, and
                         Human Evolution*
                         Riverhead Books, 2001

Lewin, Roger   *Bones of Contention:
                         Controversies in the Search for
                         Human Origins*
                         Simon & Schuster, 1987

Milton, Richard   *The Facts of Life: Shattering the
                         Myth of Darwinism*
                         Fourth Estate, 1992

Paley, William   *Natural Theology*
                         Gould and Lincoln, 1871

Ranney, Wayne   *Carving the Grand Canyon:
                         Evidence, Theories and Mystery*
                         Grand Canyon Association,
                         2005

Sagan, Carl   *Cosmos*
                         Random House, 1980

--------------   *The Demon Haunted World:
                         Science As a Candle in the
                         Darkness*
                         Random House, 1995

Stein, Ben   *"Expelled: No Intelligence
                         Allowed."*
                         Documentary on DVD, 2008

Taylor, Gordon Rattray   *The Great Evolution Mystery*
                         Harper & Row, 1983

Thaxton, Charles B.; Bradley, Walter L.; Olsen, Roger L — *The Mystery of Life's Origin: Reassessing Current Theories* Philosophical Library, 1986

Vail, Tom — *Grand Canyon: A Different View* Master Books, 2003

Ward, Peter D.; Brownlee, Donald — *Rare Earth: Why Complex Life Is Uncommon in the Universe* Copernicus, 2000

Weikart, Richard — *From Darwin to Hitler: Evolutionary Ethics, Eugenics, and Racism in Germany* Palgrave Macmillan, 2004

Wells, Jonathan — *Icons of Evolution: Science or Myth? (Why Much of What We Teach About Evolution Is Wrong)* Regnery, 2000

Werner, Carl — *Evolution: The Grand Experiment* New Leaf, 2007

About the Author

From the time he first recalls gazing at the moon as a young child in Philadelphia, Pennsylvania, John Verderame has had a sense of wonder and awe about the universe that led to an interest in, and the study of, astronomy, biology, geology, and theology, and a fascination with the space program. John has traveled extensively, including riding a bicycle across the USA and living in Italy for a number of years, where his lifelong admiration of Galileo Galilei and his willingness to "go against the tide" was solidified. He has broad work experience ranging through everything from lead singer and organist in a rock band, to health inspector, to quality control technician, to telescope manufacture, to his most recent service as a code enforcement officer for the Cody, Wyoming police department. John has been both a speaker and writer on the topics he covers in this book. He is an amateur astronomer and avid rock and book collector. The publication of this book is his latest adventure, and he could not have done it without the love and support of his wife, Laura. As noted throughout *Evolution Is Stupid!*, however, this is not about John. It's about Evolution and changing peoples' thinking.